Biomathematical Problems
in
Optimization *of*
Cancer Radiotherapy

Biomathematical Problems *in* Optimization *of* Cancer Radiotherapy

Leonid G. Hanin, Ph.D.
Researcher
Department of Mathematics
Technion-Israel Institute of Technology
Haifa, Israel

Lyudmila V. Pavlova, Ph.D.
Assistant Professor
Department of Applied Mathematics
St. Petersburg Technical University
St. Petersburg, Russia

Andrej Yu. Yakovlev, Ph.D., D.Sc.
Professor and Chair
Department of Applied Mathematics
St. Petersburg Technical University
St. Petersburg, Russia

CRC Press
Taylor & Francis Group
Boca Raton London New York

CRC Press is an imprint of the
Taylor & Francis Group, an **informa** business

CRC Press
Taylor & Francis Group
6000 Broken Sound Parkway NW, Suite 300
Boca Raton, FL 33487-2742

First issued in paperback 2019

ISBN-13: 978-0-8493-8648-0 (hbk)
ISBN-13: 978-0-367-40226-6 (pbk)

Library of Congress Card Number 93-4409
Library of Congress Cataloging-in-Publication Data

Hanin, Leonid G.
    Biomathematical problems in optimization of cancer radiotherapy /
Leonid G. Hanin, Lyudmila V. Pavlova, Andrej Yu. Yakovlev.
        p. cm.
    Includes bibliographical references and index.
    ISBN 0-8493-8648-9
    1. Cancer—Radiotherapy—Mathematical models. 2. Cancer-
-Radiotherapy—Computer simulation. 3. Radiation—Dosage-
-Mathematical models. 4. Radiation—Dosage—Computer simulation.
5. Cancer cells—Proliferation—Mathematical models. 6. Cancer
cells—Proliferation—Computer simulation. I. Pavlova, Lyudmila
V. II. Yakovlev, Andrej Yu., 1944- . III. Title.
    [DNLM: 1. Models, Biological. 2. Neoplasms—radiotherapy. QZ
269 H239b 1993]
RC271.R3H35 1993
616.99'40642'011—dc20
DNLM/DLC
for Library of Congress   93-4409
                                                                              CIP

Visit the Taylor & Francis Web site at
http://www.taylorandfrancis.com

and the CRC Press Web site at
http://www.crcpress.com

ACKNOWLEDGMENTS

Professor S. T. Rachev (University of California at Santa Barbara) and Professor L. B. Klebanov (St. Petersburg Technical University) have made very valuable contributions to the research presented in our book. We also wish to thank Professor I. L. Kruglikov (Khar'kov Institute of Medical Radiology, Ukraine) for fruitful discussions and Mrs. G. M. Ablodzej for technical assistance.

THE AUTHORS

Leonid G. Hanin, Ph.D., is currently Researcher in the Department of Mathematics at the Technion — Israel Institute of Technology. He is author and coauthor of more than 30 papers in pure and applied mathematics and of two inventions. He contributed to such diverse areas of pure mathematics as the theory of Lipschitz spaces, the Kantorovich-Rubinstein theory, algebras of smooth functions and Banach algebras, extensions and approximations of functions, and nonstationary stochastic processes.

In 1978, Dr. Hanin obtained M.Sc. degrees in Mathematics and in Mathematical Education from Leningrad University, and in 1985 obtained his Ph.D. from Steklov Mathematical Institute. From 1985 to 1990 he held Assistant Professor and Senior Research Fellow positions at the Institute of Railroad Engineering in Leningrad.

His research interests in applied mathematics relate to mathematical models in radiobiology, optimization problems, applied probability, reliability of systems, and calculation of technical devices.

Lyudmila V. Pavlova, Ph.D., is Assistant Professor at St. Petersburg Technical University's Department of Applied Mathematics, which she joined in 1989. From 1981 to 1989 she held the position of Research Associate in the Department of Biomathematics, Central Research Institute of Roentgenology and Radiology in Leningrad.

Dr. Pavlova received her M.S. degree in Applied Mathematics from Leningrad Electrotechnical Institute in 1981 and her Ph.D. in Physics and Mathematics from St. Petersburg Technical Institute in 1992.

Her interests span a wide range of biomathematical and biostatistical research, including simulation and mathematical models of biological processes, survival analysis, optimal control in cancer treatment, and development of software for biomedical studies. She has authored or coauthored 12 papers.

Andrej Yu. Yakovlev, Ph.D., D.Sc., is Professor of the Department of Applied Mathematics at St. Petersburg Technical University of which he was previously Head (from 1988 to 1992). He has contributed to such diverse fields of biomathematics and biostatistics as the kinetics of cell proliferation, simulation of normal and irradiated cell systems, modeling of RNA synthesis, survival analysis, clinical trials, competing risks theory, models of aging, carcinogenesis, and tumor recurrence, statistical methods of carcinogenic risk assessment, and optimal strategies of cancer surveillance. He is the author or coauthor of over 110 papers and 2 books: *Computer Simulation in Cell Radiobiology* (with A. V. Zorin, 1988) and *Transient Processes in Cell Proliferation Kinetics* (with N. M. Yanev, 1989) published in the *Springer Lecture Notes in Biomathematics* series. In 1978 he organized the Department of Biomathematics at the Central Research Institute of Roentgenology and Radiology in St. Petersburg. He was Head of this department from 1978 to 1988.

Dr. Yakovlev began his career as a physician in 1967. He received his Ph.D. in Biology (1973) from the Institute of Physiology of the U.S.S.R. Academy of Sciences and his D.Sc. degree in Physics and Mathematics (1981) from Moscow State University. In 1991 he was elected as a Corresponding Member and in 1992 as a Full Member of the Russian Academy of Natural Sciences. He is a member of the Council of Fellows of the Collegium Ramazzini and the European Study Group for Cell Proliferation. Dr. Yakovlev is currently a visiting professor at the Curie Institute in Paris.

TABLE OF CONTENTS

Introduction

So far, many attempts have been undertaken to increase the efficiency of cancer therapy by means of solving appropriate optimization problems, most of them based on the mathematical modeling of cell population kinetics in normal and neoplastic tissues. In particular, it is generally believed that the efficiency of cancer treatment can be substantially enhanced by an optimal scheduling of dosage with the number of fractions, their sizes, and the time intervals between them subject to optimization. The same point of view is widespread in cancer chemotherapy. To obtain an optimal schedule, use is commonly made of cell kinetics description by means of differential equations supplemented with natural criteria of optimality and restrictions on the set of admissible schedules. For comprehensive discussion of such attempts, we refer the reader to surveys by Swan,[77-80] Eisen,[14] Ivanov,[31] and others. But even in the most simplified versions of cell kinetics models, the numerical parameters involved are not accessible to direct measurement for an individual patient or cannot be estimated with sufficient precision. Therefore, modern approaches to optimal control in cancer therapy contribute mainly to the investigation into general regularities but not to the treatment of a particular patient. In our book, we have not overcome this difficulty as well. The aim of this book is much more modest — to develop some new ideas providing a more profound insight into the role of cell population heterogeneity in optimal control of fractionated irradiation of tumors with a special emphasis on the mathematical aspects of the problem. Thus, the book has almost no intersections with other works in the field, in particular, with the well-known fundamental books by Swan.

Extensive radiobiological studies have revealed a variety of factors manifesting themselves at the cellular level and responsible for the resulting effect of irradiation. Among them, the heterogeneity of cells with respect to their radiosensitivity is of prime importance because radioresistant neoplastic cells may serve as the source of tumor regrowth after treatment. The conventional approach to cell heterogeneity description presupposes distinguishing within a tumor tissue a few separate cell subpopulations, e.g., hypoxic and oxygenated cells or those residing in certain phases of the cell cycle, which differ drastically in their radiosensitivity. In a more general and regular way, this can be done by assuming the parameter responsible for variability of cell radiosensitivity to be random. In doing so, one has to proceed from a cell-survival model that induces the stochastic ordering of cells with respect to their radiosensitivity. The classical multihit-one target model of irradiated cell survival responds to this purpose perfectly. Besides, this model allows solution of optimal control problems which are quite natural for cancer radiotherapy. The formulations of such problems in this book are based on the efficiency functional defined as the difference between expected (with respect to the radiosensitivity distributions) survival probabilities for normal and neoplastic cells. A rather simple structure of this functional admits parametric formulation of optimization problems and construction of the precise upper bounds for the tumor treatment efficiency over natural classes of cell survival probabilities. The investigation into the latter problem yields some interesting by-products, e.g., the "Lipschitz" upper bound gives rise to a new family of probability metrics. It should be mentioned that the ideas employed in Chapter 2 are closely connected with the theory of extreme points and the Choquet theorems,[63] but we have never used these theoretical tools directly. Nevertheless, they could be very useful when other substantive classes of functions are considered.

Different optimization problems are discussed in Chapters 2 and 3 within the framework of the model which does not account for the time factor in fractionated irradiation. Definitely, it is very attractive to introduce this factor into the model explicitly and to include the time intervals between the consecutive fractions into the set of parameters to be optimized. However, it is clear that such a generalization of the optimal control problem would come across insurmountable (as it seems to the authors) analytical difficulties. To overcome them, we have used computer simulations (see Section 3.5). The simulation study allowed us to make some preliminary inferences on the role of transient processes in multiple-dose irradiation schemes. Chapter 1 is introductory and contains all necessary information about the "hit and target" models and their modifications. For further information and references concerning optimal control of cancer treatment, we refer the reader to the previously mentioned works by Swan.

This book is intended for mathematical readership, especially specialists in biomathematics, and graduate students who will devote themselves to biomedical problems. A background in modern calculus and fundamentals of probability theory is sufficient for understanding most of the material presented. We give detailed explanations of all statements throughout the text.

We hope this book will stimulate further research based on more sophisticated models and more refined methods.

1

Mathematical Models of Irradiated Cell Survival

1.1 INTRODUCTION

Cell radiobiology has proven to be a fertile field for the application of mathematical, especially stochastic, models. Probabilistic methods of data analysis have been inseparably linked with experimental research in this field during the whole of its history. It was the need to explain the diverse character of cell response to radiation damage that caused the ideas of "hit and target" theory. Contributions to the quantitative formulation of target principle influenced by the ideas of quantum mechanics are dated as early as the 1920s. But as it was pointed out by Turner,[86] the first significant attempt at a comprehensive theory was due to Danzer[13] in 1934. The target theory made it possible to express in mathematical form the contribution of the discrete nature of ionizing radiation and the stochastic character of radiation energy scattering on cell structures to the ultimate biological effect. Classical hit and target models will be considered in Section 1.2.

With the accumulation of experimental evidence there appeared an ever-growing need for a probabilistic description of the stochastic responses of a cell to radiation damage formed according to the hit and target principle. This has given rise to the randomized models which will be briefly discussed in Section 1.3. Section 1.5 is devoted to the problem of parametric identification of a simple randomized version of the "multihit-one target" model. In the subsequent chapters this model will serve as a basis for all considerations associated

5

with the optimal strategies in fractionated irradiation of tumors. To give a complete picture of the principal trends in the target theory, some Markovian discrete models will be described in Section 1.4.

1.2   CLASSICAL HIT AND TARGET MODELS

The contents of this section relate largely to the history of stochastic radiobiology, but it seems necessary to follow the evolution of ideas in the field. Consider a single cell exposed to an acute (short-term) irradiation. We begin with enlisting, following Turner,[86] the basic assumptions of the hit and target theory, which were explicitly formulated by Danzer.[13]

1. There exists a certain number of sensitive regions (targets) in every cell which must be hit for damage to result. The cell (organism) has M such targets.
2. The cell dies if N or more of the targets are destroyed.
3. A target is destroyed if it is "hit" by m + 1 or more radiation particles (portions of dose). Any particle may hit at most one target.
4. The probability p_i that a given radiation particle will hit the ith target is the same for all particles. This is called the Bernoulli postulate.
5. If nD is the total number of radiation particles (where n is the number of particles per unit dose and D is the dose of irradiation), then we let $nD \to +\infty$, $p_i \to 0$ in such a way that nDp_i tends to a finite limit x_iD. This is called the Poisson postulate.
6. The similar target postulate: $x_1 = x_2 = \ldots = x_M = x$.

Assumptions 1, 2, and 3 jointly may be called the threshold postulate. For detailed discussion of the grounds of the hit theory, see Reference 95. Postulates 4 and 5 imply that in the case M = 1, the distribution of the random number v of hits will be Poisson with parameter xD. From the threshold postulate, it follows that the probability of cell survival is equal to

$$S(D) = \sum_{k=0}^{m} \frac{(xD)^k}{k!} e^{-xD}, \quad m = 0, 1, \ldots \, , \qquad (2.1)$$

where x and m are considered to be free parameters of the model.

Formula 2.1 expresses the so-called multihit-one target model of the "dose-effect" relationship; it does not imply any explicit description of the recovery of cells from radiation damage. The simplest way to take the repair processes into account is to interpret the quantity xD as the expected number of primary lesions produced by the dose D. Let π be the probability for a lesion to be misrepaired after its formation (this probability is assumed to be the same for all lesions). Then, neglecting the temporal aspect of functioning of the repair system, we have

$$S(D) = \sum_{k=0}^{m} \frac{(\pi xD)^k}{k!} e^{-\pi xD}, \quad m = 0, 1, \ldots . \quad (2.1')$$

It is evident that this way of introducing the repair processes within the multihit-one target model does not change the structure of the latter. Thus, parameters π and x cannot be estimated separately from the dose-effect curve. Another simple modification of this model, which is based on the concept of discrete repair units (reparons), will be presented in the next chapter.

The straightforward interpretation of the parameter m in Formula 2.1 consists in viewing it as the critical number of radiation-induced primary lesions (assumed to be identical) a cell can bear without being killed. Indeed, the parameters of Model 2.1 implicitly contain information on heterogeneity of lesions and the processes of radiation damage repair. If we assume that primary lesions are not identical and that they are subject to repair processes, the meaning of the parameter m will become rather obscure. It will refer to a somewhat abstract quantity — the critical number of those radiation-induced lesions that are responsible for cell death. Such lesions may arise as an immediate result of irradiation, and there is every reason to identify them as the potentially lethal damage (see Reference 90 for substantiation). Another source of these lesions is their forming through the transformation from less heavy lesions that constitute the sublethal damage of a cell. Thus, variations in the value of parameter m might be thought of as an effect of sublethal damage repair operating in both actively proliferating cells and those resting ones which sustain a high degree of readiness to proliferate. In the sequel, we will adhere to this interpretation of the critical number of hits, m, while the parameter x will be interpreted as a basic characteristic of damaging process itself, i.e., as radiosensitivity in its literal sense.

Under some special experimental conditions, e.g., in experiments with a delayed explantation of cells irradiated in the stationary phase of their growth *in vitro,* potentially lethal damage repair may be observed. If in such studies the value of the parameter x is affected, this can reasonably be interpreted as a result of potentially lethal damage repair (see Formula 2.1′), the temporal aspects of this phenomenon being ignored. Of course, the simple model represented by Formula 2.1 may be instrumental only in making preliminary inferences on recovery of cells from radiation damage in a given experimental system, and a more comprehensive model incorporating explicit description of repair processes is necessary to provide a more profound insight into their organization.

Formula 2.1 may be written in the following concise form:

$$S(D) = \frac{\Gamma(m + 1, xD)}{m!}, \quad m = 0, 1, \ldots , \qquad (2.2)$$

where $\Gamma(a, u) = \int_u^\infty t^{a-1}e^{-t}dt$ is the incomplete gamma function. Considering the events of destruction or nondestruction of critical targets as the Bernoulli trials, we have for the probability of survival of a cell exposed to dose D

$$S_N(D) = \sum_{i=M-N+1}^{M} \binom{M}{i} S^i(D)[1 - S(D)]^{M-i} ,$$

where S(D) is given by Formula 2.2.

Using the notation of the incomplete beta function,

$$B_u(a, b) = \frac{\displaystyle\int_0^u t^{a-1}(1 - t)^{b-1}dt}{\displaystyle\int_0^1 t^{a-1}(1 - t)^{b-1}dt}, \quad 0 \le u \le 1 ,$$

it is possible to rewrite the formula for $S_N(D)$ as follows:

$$S_N(D) = B_{S(D)}(M - N + 1, N) .$$

Turner[86] cataloged different special variants of multihit-multitarget models in a table form. Some of the models appear to be identical in analytical form and hence indistinguishable in the course of parametric estimation of the dose-effect relationships. It is customary to call such models aliases. From our point of view, it is the artificiality of the multitarget assumption that causes the problem of aliasing in the hit theory. Actually, similar target postulate 6 makes the critical targets indistinguishable and therefore Assumptions 1 and 2 are unnecessary. The biological meaning will be unchanged if we consider only one target in a given cell instead of numerous identical targets and consequently confine ourselves to using the multihit-one target model. Nevertheless, it is the multitarget version that is most frequently used in experimental radiobiological studies.

1.3   RANDOMIZED VERSIONS OF HIT AND TARGET MODELS

Apart from the multitarget idea, there exist some other ways to generalize the multihit-one target model which seem much more natural and biologically justifiable. In particular, the generalization may be achieved through weakening the threshold postulate restrictions. Assume the parameter x to be random. Turner[86] explained the random nature of this parameter as intratarget variation in the hit probability due to purely physical causes. Within the multihit-one target framework, the parameter x may be interpreted as a general characteristic of nonhomogeneity of cells with respect to their radiosensitivity. This interpretation might imply variations of the capacity of cells to repair radiation damage (see Formula 2.1') and other fluctuating factors which exert their influence on the resultant effect of irradiation. So we consider this characteristic as a random variable (r.v.) X with the cumulative distribution function (c.d.f.) F(x).

By compounding probability 2.1, we obtain

$$S_m(D) = \sum_{k=0}^{m} \frac{D^k}{k!} \int_0^\infty x^k e^{-xD} dF(x) \quad .$$ (2.3)

Using gamma distribution for F(x), Turner expressed $S_m(D)$ via the incomplete gamma function. We will use similar formulas in Section 1.5 and Chapter 2.

Note that being a monotonically decreasing function of x, Expression 2.1 induces the stochastic ordering of cells with respect to their radiosensitivity. Another way of providing such an ordering is consideration of the critical number of hits (unrepaired/misrepaired lesions) as a discrete r.v. Let Q(k) be the probability that a cell will survive k lesions. The function Q(k) is assumed to be independent of the irradiation dose which causes the lesions. It is easy to show that in this case, the dose-effect relationship $S_x(D)$ may be represented in the form

$$S_x(D) = \sum_{k=0}^{\infty} \frac{(xD)^k e^{-xD}}{k!} Q(k) \quad . \tag{2.4}$$

The following properties of the sequence $Q(k)$ seem to be quite natural:

$$\begin{cases} Q(k) \geq Q(k + 1), \ k = 0, 1, \ldots \ , \\ Q(0) = 1, \ Q(k) \to 0 \ \text{as} \ k \to \infty \quad . \end{cases} \tag{2.5}$$

In the paper by Clifford,[9] a class of nonthreshold models (2.4) with $Q(k)$ satisfying conditions (2.5) was called Class A. This class contains the conventional multihit model as a special case with

$$Q(k) = \begin{cases} 1, & k \leq m \ ; \\ 0, & k > m \quad . \end{cases} \tag{2.6}$$

As it was proven by Clifford, the parameter x and the sequence $Q(k)$ are not jointly identifiable. This is established by the following statement.

Statement 1 — If $S_x(D)$ is a function of the class A with given x and $Q(k)$, then for each $y \geq x$, there exists a sequence $Q^*(k)$, satisfying 2.5, such that

$$\sum_{k=0}^{\infty} \frac{(xD)^k e^{-xD}}{k!} Q(k) = \sum_{k=0}^{\infty} \frac{(yD)^k e^{-yD}}{k!} Q^*(k) \quad .$$

The following relationship between $Q(k)$ and $Q^*(k)$ holds

$$Q^*(k) = \sum_{n=0}^{k} Q(n) \binom{k}{n} \left(\frac{x}{y}\right)^n \left(1 - \frac{x}{y}\right)^{k-n} \quad .$$

Another important result obtained by Clifford[9] is formulated as follows.

Statement 2 — If S(D) is a function in Class A, then there is a least value y_0 of y for which a representation in this class is possible; the parameter y is restricted by the inequality

$$y \geq \max \left\{ -S'(0), \; \frac{\int_0^\infty S(u)du}{2\int_0^\infty uS(u)du - \left(\int_0^\infty S(u)du\right)^2} \right\} . \qquad (2.7)$$

In the special case of the threshold model given by 2.4 and 2.6, this estimate has the form $y \geq x$, i.e., the parameter x, for which $S_x(D)$ has the multihit representation, is defined by the relation

$$x = \frac{\int_0^\infty S(u)du}{2\int_0^\infty uS(u)du - \left(\int_0^\infty S(u)du\right)^2} ,$$

and the critical number of hits is given by

$$m = x\int_0^\infty S(u)du - 1 . \qquad (2.8)$$

This result can be interpreted as follows. Cell death probability, $G(D) = 1 - S(D)$, is a nondecreasing function of the absorbed dose, D, satisfying the conditions $G(0) = 0$, $G(+\infty) = 1$. According to Turner,[86] one can define a probability space $(\Omega, \mathcal{F}, P)$ and a r.v. Y (called stochastic inactivating dose) on it such that the c.d.f. for Y coincides with G, i.e., $P(Y \geq D) = S(D)$. The inequality 2.7 means that the rate of the Poisson process of the lesion formation per unit dose cannot be less than $\mathbb{E}\{Y\}/\mathrm{Var}\{Y\}$, where $\mathbb{E}\{Y\}$ is the expectation of Y, and $\mathrm{Var}\{Y\}$ is its variance. Thus,

$$y \geq \max \left[-S'(0), \; \frac{\mathbb{E}\{Y\}}{\mathrm{Var}\{Y\}} \right] . \qquad (2.9)$$

Recalling the model given by Expression 2.3, one can see that it is uniquely determined by the distribution function F.

The following theorem is proven by Klebanov.

Theorem — Assume that distribution functions $F_i(x)$, $i = 1, 2$, vanish for $x < 0$. Let $S_{m_i}^{(i)}(D)$ be the survival probabilities given by

$$S_{m_i}^{(i)}(D) = \sum_{k=0}^{m_i} \frac{D^k}{k!} \int_0^\infty x^k e^{-xD} dF_i(x), \quad i = 1, 2 \quad .$$

1. If $m_1 = m_2 = m$ (m is known) and $S_m^{(1)}(D) = S_m^{(2)}(D)$ for all $D \geq 0$, then

$$F_1(x) = F_2(x) \quad \text{for all } x \geq 0 \quad .$$

2. If $F_i(x)$ are gamma distributions with unknown parameters α_i, β_i, $i = 1, 2$, and $S_{m_1}^{(1)}(D) = S_{m_2}^{(2)}$ for all $D \geq 0$, then

$$\alpha_1 = \alpha_2, \ \beta_1 = \beta_2, \ m_1 = m_2 \quad .$$

For further discussion of this problem, see Section 1.5.

More complicated models of hit type allow description of the enzymatic repair of radiation damage. But in a more clear way, appropriate generalizations may be obtained within the framework of Markov processes. Such an approach was partially used by Clifford in the above-mentioned paper. Markov models of radiation cell death will be briefly reviewed in the next section.

1.4 MARKOV FORMULATIONS OF HIT MODELS

Accumulation of lesions in the course of irradiation is usually considered as a continuous-in-time Markov stochastic process $\{X(\tau), \tau > 0\}$ defined on the discrete state space $x = \mathbb{Z}_+$. The symbol $\mathbb{Z}_+$ is used hereafter for the set of nonnegative integers (numbers of lesions). It is assumed that $X(\tau)$ is a locally regular process, i.e., the mean time of the cell residence in any state $n \in \mathbb{Z}_+$ differs from zero. Thus, the transition probabilities

$$P(n, \tau | k, \tau_0) = P\{X(\tau) = n | x(\tau_0) = k\}, \ \tau > \tau_0 \quad ,$$

satisfy the forward Kolmogorov equation. In the case of single exposure to ionizing radiation, variable τ may be considered as the duration of irradiation or the absorbed dose because of the assumed linear dependence, $\tau = cD$, between these values in the course of irradiation.

In the classical scheme of Blau-Altenburger,[95] it was supposed that $X(\tau)$ is a pure birth process, and that for each event of elementary interaction of the radiation with the cell substance, the number of lesions changes no more than in 1. This model yields Expression 2.3 when for the cell death it is necessary to inactivate N identical targets and when a target is destroyed if it is hit by $m + 1$ or more radiation particles.

Kruglikov[42] considered the case when the probability of transition $n \rightarrow n + 1$ equals $a\Delta\tau + o(\Delta\tau)$, and the probability of formation of two lesions in time interval $\Delta\tau$ also differs from zero and equals $b\Delta\tau + o(\Delta\tau)$. He demonstrated that, for small values of b, the first moments of lesion distribution in a cell are given by the approximate expressions

$$\mathbb{E}\{X(\tau)\} \sim aD + 2bD \quad,$$

$$\text{Var}\{X(\tau)\} \sim aD + 4bD(1 - bD) \quad.$$

Opatowski[55-57] used for the description of radiation cell damage development the birth-and-death scheme, where the rates of transitions $n \rightarrow n + 1$ and $n \rightarrow n - 1$ are λ_n and μ_n, respectively; this means that both damaging and repairing mechanisms are premised. It has to be noted that in this model, only the "rapid" repair is presumed, i.e., repair occurs only during the irradiation. The state $n = r$ is absorbing, i.e., repairing is impossible, if the state r is reached. In this case

$$S(\tau) = \sum_{n=0}^{r-1} P(n, \tau) = 1 - A_r\tau^r \left[\frac{1}{r!} + \sum_{j=1}^{\infty} \frac{\alpha_{rj}\tau^j}{(r + j)!}\right] \quad,$$

where

$$A_i = \begin{cases} 1, & i = 0 \ ; \\ \prod_{k=1}^{i} \lambda_k, & i = \overline{1, r} \ , \end{cases}$$

and α_{rj} depend on the coefficients λ_n and μ_n only. For low radiation doses, the cell death probability, $G(\tau) = 1 - S(\tau)$, is given by

$$\ln G(\tau) \sim r\ln\tau + \ln A_r - \ln r!, \quad \tau \to 0 \quad .$$

Hence, one can obtain the asymptotic number of intermediate states of cell damage before its death,

$$r \sim \tau \frac{d}{d\tau} \ln G(\tau), \quad \tau \to 0 \quad .$$

Bharucha-Reid and Landau[7] used the scheme of stochastic wandering for describing the radiation damaging. They supposed that the quantum of radiation or a charged particle hitting the sensitive volume of a cell (or some control molecule) leads to development of the chain reactions of depolymerization of some macromolecules. The process of damage development was described by the finite Markov chain with $r + 1$ states. For the survival of n cells among N irradiated cells, the following formula was obtained:

$$S_{n,N} = \binom{N}{n}(1 - 2^{1-r})^n(2^{1-r})^{N-n} \quad .$$

Rubanovich[70] considered the birth-and-death model given that the survival function S(D) has the asymptotics

$$\ln S(D) \sim \alpha - \beta D, \quad \alpha, \beta > 0, \quad D \to +\infty \quad .$$

For the r.v. Q, which is the number of states that were occupied by a cell before its inactivation, he has obtained the estimate

$$\mathbb{E}\{Q\} \geq \max\left\{\frac{\mathbb{E}^2\{Y\}}{\mathrm{Var}\{Y\}}, \beta\mathbb{E}\{Y\}\right\} \tag{3.1}$$

(recall that Y is the stochastic inactivating dose).

It has to be noted that the first of these estimates was obtained by Hug and Kellerer,[30] and it is worth comparing 3.1 with Clifford's estimate 2.9. Rubanovich[70] considered also the process that is strictly staged with respect to lesion formation, i.e., it is characterized by the relation

$$P\{X(\tau + \Delta) = n | X(\tau) = k\} \sim \delta_{n,k+1}, \quad \Delta > 0 \quad ,$$

where $\delta_{n,k}$ is, as usual, the Kroneker symbol defined by $\delta_{n,k} = 0$ when $n \neq k$, and by $\delta_{n,k} = 1$ when $n = k$. For the probability l_i of the intermediate state i, he obtained the following inequality:

$$l_i \leq \min \left\{ \frac{1 + \dfrac{1}{\beta \mathbb{E}\{Y\}}}{1 + \dfrac{\text{Var}\{Y\}}{\mathbb{E}^2\{Y\}}}, \quad 1 \right\} .$$

The estimate is sharp when $\beta \, \text{Var}\{Y\} \geq \mathbb{E}\{Y\}$.

Rubanovich[71] obtained the upper bound for the probability that at least once the cell will return in the initial state, $X = 0$, before its death:

$$P\{X(\tau) = 0 | X(0) = 0\} \leq \left[\frac{4(\beta \mathbb{E}\{Y\} - 1)}{\beta^2 (\mathbb{E}^2\{Y\} + \text{Var}\{Y\}) - 2} \right]^2 .$$

Time-varying distribution of the lesion number influenced by proliferation and death of irradiated cells was studied in the book by Yakovlev and Yanev.[89] Consider a population of irradiated cells that divide and die independently of each other. We focus our attention on the reproductive type of radiation-induced cell death. Let $t = 0$ be the end of irradiation time. In accordance with the hit and target model, the distribution of the number of lesions in a cell sampled immediately after irradiation is expected to be Poisson. Obviously, for $t > 0$, this distribution does not remain the same; it undergoes postirradiation changes which are only partially caused by the recovery of cells from radiation damage. The processes of multiplication and death of irradiated cells also contribute to these changes because when a cell dies, all its lesions disappear, and when a cell divides in two, each lesion from the mother cell is distributed with certain probabilities between the daughter cells. The problem is describing the time-varying distribution of the lesion number in the presence of cell multiplication and cell death. It is possible to solve the problem by neglecting the postirradiation repair.

Intracellular lesions induced by irradiation may be interpreted as discrete marks attached to a cell. As a rule, mathematical models of cell population kinetics determine the fate of a cell at the end of mitotic cycle. With probability p, a cell produces two viable descendants, and with probability $1 - p$, it dies (the reproductive type of death).

Considering an irradiated cell population from this point of view, it is natural to assume that the value of cell division probability, p, depends on the number, n, of such marks. Yakovlev and Yanev[89] specified the dependence

$$p_n = pa^n, \ 0 < a \leq 1, \ 0 < p < 1, \ n \in \mathbb{Z}_+ \ , \tag{3.2}$$

that fits well the radiobiological observations.

Assume evolution of irradiated cell population to be governed by the process of binary splitting and death of cells only. At the initial moment $t = 0$ in each cell of a given population, a random number of lesions is induced, and the distribution of this number is assumed to be Poisson with parameter Θ:

$$\Pi_j(0) = \frac{\Theta^j}{j!} e^{-\Theta}, \ j \in \mathbb{Z}_+ \ . \tag{3.3}$$

Then we accept the following natural assumption: when a cell dies its lesions disappear; when a cell divides in two, each lesion from the mother cell is distributed independently and with probability $1/2$ between the daughter cells. Introduce a stochastic process $Z_j(t)$ to represent a number of cells bearing exactly j lesions at the moment t, and define the distribution $\Pi_j(t)$ as the ratio of the expectations of the processes $Z_j(t)$ and $\sum_{j=0}^{\infty} Z_j(t)$. That is,

$$\Pi_j(t) = \frac{N_j(t)}{N(t)} \ , \tag{3.4}$$

where $N_j(t) = \mathbb{E}\{Z_j(t)\}$, $N(t) = \mathbb{E}\left\{\sum_{j=0}^{\infty} Z_j(t)\right\}$.

Williams[88] studied cell population kinetics within the model of a linear nonhomogeneous birth-and-death process given the initial number of marks that was distributed according to 3.3, and derived the following expression for distribution 3.4:

$$\Pi_j(t) = \frac{\Theta^j}{j!} \exp[-2\Lambda(t)] \sum_{n=0}^{\infty} (-1)^n \frac{\Theta^n}{n!} \exp[2^{1-n-j}\Lambda(t)] \ ,$$

where $\Lambda(t) = \int_0^t \lambda(u)du$, with $\lambda(t)$ being the rate of cell multiplication.

In the case of homogeneous birth-and death process, $\Lambda(t) = \lambda t$, and the above-mentioned distribution is evidently simplified to

$$\Pi_j(t) = \frac{\Theta^j}{j!} \exp[-2\lambda t] \sum_{n=0}^{\infty} (-1)^n \frac{\Theta^n}{n!} \exp[2^{1-n-j}\lambda t] \quad . \tag{3.5}$$

Yanev and Yakovlev[91] considered the same problem within the Bellman-Harris age-dependent branching process model defined by the d.f. G(t) for the mitotic cycle duration, and the probability generating function h(s) for the number v of the cell progeny,

$$h(s) = \mathbb{E}\{s^v\} = ps^2 + 1 - p \quad ,$$

where p is the probability of successful mitotic division and $1 - p$ is the probability of the reproductive death of a cell. For the cell population synchronized at zero age point in the mitotic cycle, Yanev and Yakovlev obtained the following distribution of marks:

$$\Pi_j(t) = \frac{\Theta^j}{j!} \frac{\sum_{k=0}^{\infty} (2^{1-j}p)^k \exp\left(-\frac{\Theta}{2^k}\right)[\overline{G} * G_k](t)}{\sum_{k=0}^{\infty} (2p)^k[\overline{G} * G_k](t)} \quad , \tag{3.6}$$

where $\overline{G} = 1 - G$ and G_k is the k-fold convolution of the d.f. G. The authors have also presented generalizations of the expression for $\Pi_j(t)$ covering the cases of induced cell proliferation and exponentially growing cell population.

The regular Bellman-Harris process in the case $G(t) = 1 - e^{-bt}$ degenerates into the Markov branching process. In turn, by introducing the death rate, μ, and allowing

$$b = \lambda + \mu, \quad p = \frac{\lambda}{\lambda + \mu} \quad ,$$

one may revert to the model of linear homogeneous birth-and-death process and transform 3.6 into Williams' result (3.5).

In the monograph by Yakovlev and Yanev,[89] it was shown that the distribution $\Pi_j(t)$ has the form

$$\Pi_j(t) = \frac{\Theta^j}{j!} \frac{\sum\limits_{k=0}^{\infty} \left(2p\left(\frac{a}{2}\right)^j\right)^k \exp\left\{-\frac{\Theta}{1-a/2}\left(\frac{a}{2}\right)^{k+1}\right\}[\overline{G} * G_k](t)}{\sum\limits_{k=0}^{\infty} (2p)^k \exp\left\{\frac{\Theta(1-a)}{1-a/2}\left(\frac{a}{2}\right)^k\right\}[\overline{G} * G_k](t)}$$

if dependence 3.2 is taken into account. In the special case a = 1, the latter expression degenerates into 3.6.

It is impossible to observe the temporal pattern of the number of radiation-induced intracellular lesions in a direct experiment. To measure the biological effect of irradiation, the clonogenic capacity (survival) of cells is commonly used. This experimental indicator may be considered as an estimator for the probability of extinction of the process $Z(t) = \sum\limits_{j=0}^{\infty} Z_j(t)$. If the process $Z(t)$ starts from one cell bearing a random number of lesions distributed in accordance with 3.3, then the extinction probability $r(t; \Theta) = P\{Z(t) = 0\}$ satisfies the integral equation

$$r(t; \Theta) =$$

$$G(t)[1 - pe^{-\Theta(1-a)}] + pe^{-\Theta(1-a)} \int_0^t r^2\left(t - u; \frac{a\Theta}{2}\right) dG(u) \quad , \quad (3.7)$$

and its limit value for large t can be identified as the expected clonogenic capacity of cells.

Yakovlev and Yanev[89] pointed out the problem consisting in examination of the asymptotic behavior of the extinction probability $r(t; \Theta)$ as $t \to +\infty$ on the basis of Equation 3.7. Another problem worth mentioning is obtaining an analytical expression for the distribution $\Pi_j(t)$, allowing for the repair process which results in disappearance of some lesions before the cell division.

Naturally, discrete Markovian models allow us to describe the post-irradiation recovery processes as well. For this purpose, Tobias et al.[84] and Tobias[83] considered a nonlinear model, namely, the so-called repair-misrepair (RMR) model. They assumed that irradiation is instantaneous at t = 0, and described the processes of linear repair and quadratic misrepair of lesions as follows:

$$\mathbb{E}\{X(0)\} = cD \quad ,$$

$$\frac{d}{dt}\mathbb{E}\{X(t)\} = -\nu_1\mathbb{E}\{X(t)\} - \nu_2\mathbb{E}^2\{X(t)\} \quad , \qquad (3.8)$$

where c, v_1, and v_2 are nonnegative constants, D is the total irradiation dose, and X(t) is the number of lesions at a moment t > 0 elapsed from the irradiation. The solution of 3.8 is

$$\mathbb{E}\{X(t)\} = [((cD)^{-1} + v)e^{v_1 t} - v]^{-1}$$

with $v = v_2/v_1$. The initial lesion distribution is supposed to be the Poisson one, and the lesions resulting from the quadratic misrepairs are assumed to be lethal. A cell survives if it has no lethal lesions, and this leads to the formula

$$S(D) = e^{-cD} \left(1 + \frac{cD}{v}\right)^v \qquad (3.9)$$

(for proof and further details, see Section 1.7). For integer v, one can represent 3.9 in the form of series 2.4 with

$$Q(k) = \frac{v(v - 1) \ldots (v - k + 1)}{v^k}, \quad k \in \mathbb{N}; \quad Q(0) = 1 \quad .$$

Albright and Tobias[2] used the discrete branching Markov process for description of the time-dependent cell repair after irradiation. In another work,[1] Albright also formulated the RMR model as a Markov process, i.e., as a sequence of discrete repair steps occurring at random times. His model describes the time evolution of the probability distribution for the number of lesions in a cell. Albright's model used the same biological hypotheses as the original version of the RMR model with two approximations deleted. These approximations are the neglect of the effect of random fluctuations on the rate of lesion repair and the assumption that the final number of unrepaired and lethally misrepaired lesions has a Poisson distribution. Albright[1] and Brenner[8] also considered the case when the number of primary lesions produced in one event and the amount of specific energy imparted are stochastically dependent.

Albright[1] used a two-dimensional birth-and-death process {X(t), Y(t), t > 0} to describe the evolution of two types of lesions after irradiation. He introduced the probability P(n, k, t|D) = $P_{n,k}(t)$ that a cell has X(t) = n primary (unrepaired) lesions and Y(t) = k secondary (misrepaired) lesions, n, k $\in \mathbb{Z}_+$. Unrepaired or misrepaired lesions are lethal. Albright considered two types of repair: a linear repair that involves only one lesion, and a quadratic repair that involves the interaction of two lesions. A fraction $(1 - \phi)$ of linear repairs and a fraction $(1 - \psi)$ of quadratic repairs are assumed to be lethal misrepairs. Quadratic repair gives one unrepaired and one repaired or misrepaired lesion. The following forward Kolmogorov equation for $P(n, k, t|D)$ holds

$$\frac{dP_{n,k}(t|D)}{dt} = -\alpha_n P_{n,k}(t|D) + \beta_{n+1} P_{n+1,k}(t|D) +$$

$$(\alpha_{n+1} - \beta_{n+1}) P_{n+1,k-1}(t|D) \quad .$$

Here $\alpha_n \Delta t$ is the probability of linear or quadratic repair in time Δt, and $\beta_n \Delta t$ is the probability of viable linear or quadratic repair in time Δt:

$$\alpha_n = \lambda_1 n + \lambda_2 n(n - 1) \quad ,$$

$$\beta_n = \phi \lambda_1 n + \psi \lambda_2 n(n - 1) \quad ,$$

where λ_1 and λ_2 are nonnegative constants.

It was shown that the equation for $E\{X(t)\}$ has the form

$$\frac{dE\{X(t)\}}{dt} = -\lambda_1 E\{X(t)\} - \lambda_2 E^2\{X(t)\} -$$

$$\lambda_2[Var\{X(t)\} - E\{X(t)\}], \quad t \geq 0 \quad , \tag{3.10}$$

with the initial condition

$$E\{X(0)\} = cD \quad .$$

Equation 3.10 is equivalent to 3.8 when $Var\{X\} = E\{X\}$, in particular, for the Poisson distribution of $X(t)$.

Obaturov et al.[52,53] took into account the following processes after irradiation: linear repair without formation of the secondary lesions with the probability αn and quadratic interaction with formation of the lethal for a cell exchanged aberrations with the probability $\beta n(n - 1)$. In this case, the forward Kolmogorov equation has the form

$$\frac{dP_{n,k}(t)}{dt} = \beta(n + 1)(n + 2)P_{n+2,k-1}(t) + \alpha(n + 1)P_{n+1,k}(t) -$$

$$[\beta n(n - 1) + \alpha n]P_{n,k}(t), \quad t \geq 0 \quad ,$$

where n, k $\in \mathbb{Z}_+$ and α, $\beta > 0$. The corresponding equations for $\mathbb{E}\{X(t)\}$ and $\mathbb{E}\{Y(t)\}$ in this model are

$$\frac{d}{dt} \mathbb{E}\{X(t)\} = -\alpha\mathbb{E}\{X(t)\} - 2\beta[\mathrm{Var}\{X(t)\} + \mathbb{E}^2\{X(t)\} - \mathbb{E}\{X(t)\}],$$

$$\frac{d}{dt} \mathbb{E}\{Y(t)\} = \beta[\mathrm{Var}\{X(t)\} + \mathbb{E}^2\{X(t)\} - \mathbb{E}\{X(t)\}] \ . \qquad (3.11)$$

These equations were studied in References 6 and 11. It is worth noting that the models by Obaturov et al.[52,53] and Albright[1] are indistinguishable as far as expectations are concerned.

The deterministic-type model by Curtis[11] is based on the idea that repairable lesions can produce irreparable ones because of the process of damage fixation which may be interpreted as misrepair as well. The transformations of lesions are described by nonlinear differential equations similar to 3.10 and 3.11. The approach was generalized by Thames[82] to include the effects of dose fractionation and low-dose rate (see Section 7 for more details).

Some further developments in the repair dynamics description can be found in the works of Payne and Garrett.[60,61] In the refined version of their model, Garrett and Payne[18] proposed the Markov model with continuous time and three discrete states of a target: A — absence of damage, B — presence of repairable damage, and C — presence of irreparable damage. The principal element of the model is the concept of "fixation time" which specifies that all repairable damage (state B) becomes irreparable (state C) unless its repair is effected within a certain time interval called fixation time. Cells suffering greater than a given number of lesions in state C die the reproductive type of death.

Another Markovian model which seems to be rather close to that of Payne and Garrett was developed by Neyman and Puri.[50,51] The reader may refer to Puri[67] for a brief discussion of this model. The most impressing feature of the model is that it presumes not only cell killing but also neoplastic transformation as possible outcomes of the radiation effect. It is competition between the two risks that allows prediction of the observed pattern of cell transformation *in vitro* for different doses of irradiation.[67]

Kapul'tcevich and Korogodin,[36,37] Kappos and Pohlit,[35] Laurie et al.,[47] Goodhead et al.,[21,22] and others considered different phenomenological models for the description of evolution of the expected number of lesions. Further progress in stochastic description of irradiated cell survival within the framework of discrete Markov processes was due to works by Yang and Swenberg.[92,93] They assumed that each primary radiation particle generates a random number, M, of spurs. Each spur generates a potentially lethal lesion with probability π_1 and a lethal lesion with probability π_2. With probability $1 - \pi_1 - \pi_2$, the spur has no effect on the cell. The joint distribution of the number of potentially lethal lesions, U_1, and the number of lethal lesions, U_2, is trinomial. The evolution of the cell during and after irradiation is described by a vector-valued stochastic process (X_t, Y_t, Z_t), where X_t is the number of potentially lethal lesions, Y_t is the number of lesions producing neoplastic transformation of a cell, and Z_t is the number of lethal lesions at time t. The following transitions are allowed in their scheme: $(X_t, Y_t, Z_t) \rightarrow (X_t - 1, Y_t, Z_t)$ with the probability $\alpha X_t \Delta t + o(\Delta t)$ (correct repair); $\rightarrow (X_t - 1, Y_t + 1, Z_t)$ with the probability $\beta X_t \Delta t + o(\Delta t)$ (misrepair in neoplastic transformation); $\rightarrow (X_t - 1, Y_t, Z_t + 1)$ with the probability $\gamma X_t \Delta t + o(\Delta t)$ (misrepair in inactivation); $\rightarrow (X_t, Y_t, Z_t + 1)$ with the probability $\delta \Delta t + o(\Delta t)$ (cell death or inactivation unrelated to radiation); $\rightarrow (X_t + u_1, Y_t, Z_t + u_2)$ with the probability $\lambda A \Delta t P\{U_1 = u_1, U_2 = u_2\} + o(\Delta t)$ for all u_1 and u_2 (initiation of new lesions), where A is the size of the sensitive region. In this model, the survival probability at low-dose rates is an exponential function of the dose with the decay constant independent of the dose rate.

Continuous Markov processes and stochastic differential equations were applied for the analysis of the interphase death of cells exposed to radiation.[40-46] For this radiobiological phenomenon, it is reasonable to consider the number of lesions to be very large. Moreover, not all initial lesions in a cell are biologically identical[59,87] and to the extent that they reflect a continuum of possible levels of damage, it seems plausible to use noninteger values of $X(\tau)$.[72]

Sachs and Hlatky[72] and Sachs et al.[73] incorporated stochastic dose-rate effects in Markov radiation cell survival models. Using stochastic differential equations and the Monte Carlo method to take into account the stochastic effects, they calculated the dose-survival relationships in a number of current cell survival models with quadratic misrepair and saturable repair enzyme systems. They found that in the limit of low-dose rates, neglecting the stochastic nature of specific energy rates frequently brings misleading results by overestimating the surviving fraction. In the opposite limit of acute (short-term) irradiation at high-dose rates, neglecting the fluctuations usually underestimates the surviving fraction.

It has to be emphasized that the homogeneous birth and death model represents a quite simplified description of biological processes of cell recovery from radiation damage. Goodhead[21] adduced experimental facts supporting the following premises which might serve as a basis for mathematical modeling of radiation effect upon a cell:

- Radiation damage occurs at short distances from the sites of absorption of small amounts of energy — the predominant role in producing damage being played by the single-track mechanism.
- Number of lesions is proportional to the radiation dose.
- Damage can be reduced by the intracellular repair system whose efficiency drops with the increase of the radiation dose.

Alper[4] also came to the conclusion that radiation damage occurs in accordance with the one-hit mechanism and that the relatively low sensitivity of cells in the event of small doses is associated not with the multiplicity of targets or the interaction between separate lesions but with the existence of a certain biochemical mechanism of repair. Alper stressed that high doses of irradiation suppress the repair mechanism's functioning, apparently due to the exhaustion of the biochemical factors (repair enzymes) involved in it. The author noted that only in the case of exponential survival curve, there is a reason to believe that either the mechanism does not function at all or it maintains equal efficiency over the entire range of doses. According to Alper,[4] this idea was first set forth by Powers.[66] Then Sinclair[74] advanced the hypothesis of the existence of a certain factor, whose concentration fluctuations determine variations in the magnitude of the survival curve shoulder, as the cells traverse the mitotic cycle.

Several mathematical models have been proposed to formalize the idea of the existence of a saturable repair system.[3,22,23,35,47,58] These models differ from each other mainly in the analytical form of relationship between repair efficiency and irradiation dose. A similar approach was also developed for describing the effect of ultraviolet irradiation of cells.[20,27,28] According to the Haynes model,[27] the number of repairable lesions approaches saturation with the increase of the irradiation dose. In his later work,[28] Haynes modified the model by introducing the extreme form of relationship between repair efficiency and dose. Similar form of this relationship was also used by Yakovlev and Zorin[90] within a computer simulation model intended to reproduce various phenomena of cell radiobiology. More mathematical work is to be done to formalize the above considerations from the standpoint of strict probabilistic methods.

1.5 IDENTIFICATION OF THE RANDOMIZED MULTIHIT-ONE TARGET MODEL

As was pointed out in Section 1.3, the multihit-one target model of irradiated cell survival given by Formula 2.1 provides the stochastic ordering of cells with respect to their radiosensitivity, if the parameter x is treated as a random variable X with some specified distribution function F(x). This consideration allows construction of various randomized versions of model 2.1 by applying the formula

$$S_m(D) = \sum_{k=0}^{m} \frac{D^k}{k!} \int_0^\infty x^k e^{-xD} dF(x) \quad .$$

In what follows, we shall always assume that the critical number of hits m is nonrandom, i.e., it takes the same value for all cells in a given population.

The distribution function F(x) is not observable, and it depends on the vector Θ of unknown parameters, i.e., $F(x) = F(x; \Theta)$. Thus, introducing a randomized version of the cell survival model, one meets with an increased number of parameters to be estimated from experimental observations. The main question is whether the model under study retains its identifiability after the application of randomization procedure. The positive answer to this question (see Section 1.3) means that an appropriate estimate for the radiosensitivity distribution may be obtained from data of the dose-effect type. There is still another question: how good is a particular estimation procedure? It is evident that a special study is needed for every parametric formulation of this problem. Presented below are the results of such a study carried out for the multihit-one target model given by Formula 2.1 with the gamma-distributed parameter x and nonrandom parameter m. In this case, the survival probability can be represented in the form

$$S(D; \alpha, \beta, m) = \left(\frac{\beta}{\beta + D}\right)^\alpha \sum_{k=0}^{m} \left(\frac{D}{\beta + D}\right)^k \frac{\Gamma(k + \alpha)}{k!\Gamma(\alpha)} \quad , \quad (5.1)$$

where $\Gamma(u)$ is the gamma function, α and β are the shape and scale parameters of the involved gamma distribution, respectively.

First of all, it is necessary to make sure that, using a given estimation procedure, we obtain appropriate estimates of the parameters α, β, m, if the observed survival frequencies at various dose values follow exactly the theoretical relationship given by 5.1. This can be done with the aid of computer simulations.

In our study, simulation experiments were conducted as follows. Using Formula 5.1, the values of S(D) were computed for the prescribed values of dose $D = D_i$, $i = 1, \ldots, r$, and for the model parameters $\alpha = \alpha^*$, $\beta = \beta^*$, $m = m^*$. A random variable ξ uniformly distributed on the interval (0, 1) was generated for each cell in the simulated population. A cell was considered as having survived if the value of ξ was greater than $S(D_i; \alpha^*, \beta^*, m^*)$, otherwise it was considered dead. This experiment was repeated L times for every cell from the given population of size N and for every dose value D_i, $i = 1, \ldots, r$. As a result, realizations were obtained of each random variable v_{ij}, $i = 1, \ldots, r$; $j = 1, \ldots, L$, representing the number of cells that had survived the dose D_i in the jth sample.

The least squares estimates (LSE) for the parameters α, β, and m were obtained from samples v_{ij} by minimizing the function

$$\sum_{i=1}^{r} [S(D_i; \alpha, \beta, m) - \overline{S}(D_i)]^2 \ , \tag{5.2}$$

where $\overline{S}(D_i) = \dfrac{1}{L} \sum_{j=1}^{L} S_j(D_i)$, $S_j(D_i) = \dfrac{v_{ij}}{N}$.

The search for the minimum of 5.2 with respect to α, β, and m was carried out numerically by computer using the flexible simplex method described in the book by Himmelblau.[29] Then the LSEs $\hat{\alpha}$, $\hat{\beta}$, and $\hat{m}$ were used as initial parameter values in the computational algorithm of searching for the maximum likelihood estimates (MLE) $\hat{\alpha}'$, $\hat{\beta}'$, and $\hat{m}'$ from the same sample. The log-likelihood function for this type of data has the form

$$\sum_{i=1}^{r} \sum_{j=1}^{L} \left\{ \log \binom{N}{v_{ij}} + v_{ij} \log S(D_i) + (N - v_{ij}) \log[1 - S(D_i)] \right\} \ ,$$

where $S(D_i) = S(D_i; \alpha, \beta, m)$ is given by 5.1.

Shown in Table 1 are the results of one set of simulation experiments conducted with the following values of the true parameters of the model: $\alpha^* = 16$, $\beta^* = 40$, and $m^* = 5$. These results demonstrate stability of the LSE values with respect to the choice of the initial

Table 1

Influence of Initial Approximation on Estimates of the Parameters α, β, m

Initial Approximation			L S E			M L E		
α_0	β_0	m_0	$\hat{\alpha}$	$\hat{\beta}$	$\hat{m}$	$\hat{\alpha}'$	$\hat{\beta}'$	$\hat{m}'$
20	50	10	18.9	46.1	5	18.7	46.1	5
20	50	20	15.1	31.2	6	16.3	39.2	5
10	50	20	16.7	47.4	4	16.6	47.1	4
10	50	10	18.6	45.3	5	18.5	45.0	5
10	30	10	18.7	45.6	5	18.6	45.3	5
10	70	10	18.7	45.6	5	18.6	45.3	5

Note: Sample size: N = 20; the number of replications: L = 10; true values of the parameters: $\alpha^* = 16$, $\beta^* = 40$, $m^* = 5$.

approximation. By comparing the estimates of the parameters, α, β, m with their true values, one may see that the least squares method provides a good estimation of the unobservable characteristics of the radiosensitivity distribution and of the critical number of hits from the data of the dose-effect type. Subsequent application of the maximum likelihood method does not lead to a marked improvement of the estimation procedure. Table 2 shows that the accuracy of estimates increases with the growth of the number of replications, i.e., the procedure gives consistent estimators for the parameters under study. In a more visual way, the above conclusions can be substantiated by handling not the parameters α and β themselves but the expected (mean) value $\mu = \alpha/\beta$ of the radiosensitivity distribution and its variance σ, $\sigma^2 = \alpha/\beta^2$ (Table 3). Figure 1 shows goodness of fit provided by the parametric estimate $S(D; \hat{\alpha}', \hat{\beta}', \hat{m}')$.

Table 2

Estimates of the Parameters α, β, m for Varying Replication Numbers

	L S E			M L E		
L	$\hat{\alpha}$	$\hat{\beta}$	$\hat{m}$	$\hat{\alpha}'$	$\hat{\beta}'$	$\hat{m}'$
10	17.4	49.3	4	17.4	49.2	4
100	19.7	43.8	6	20.6	45.8	6
1000	18.5	46.9	5	18.4	46.7	5

Note: True parameter values: $\alpha^* = 16$, $\beta^* = 40$, $m^* = 5$; initial values: $\alpha_0 = 10$, $\beta_0 = 50$, $m_0 = 10$; sample size: N = 20.

Table 3

Estimates of the Parameters $\mu = \alpha/\beta$ and $\sigma = \sqrt{\alpha}/\beta$ for Varying Replication Numbers

	LSE		MLE	
L	$\hat{\mu}$	$\hat{\sigma}$	$\hat{\mu}'$	$\hat{\sigma}'$
10	0.353	0.085	0.354	0.085
100	0.450	0.101	0.450	0.099
1000	0.394	0.092	0.394	0.092

Note: True parameter values: $\mu^* = 0.4$, $\sigma^* = 0.1$.

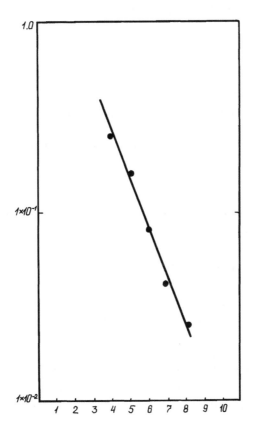

Figure 1

"Goodness of fit" provided by the parametric estimate $S(D; \hat{\alpha}', \hat{\beta}', \hat{m}')$.

Inferences made from simulation experiments refer to the identification procedure itself, but they do not allow evaluation of the model validity. To verify the model of irradiated cell survival, appropriate experimental data must be analyzed in a similar way.

Shown in Figures 2 and 3 are the experimental data on the dose-effect relationships given in the paper by Kim et al.[38] for the irradiated cultures of HeLa and 3T3 cells. Using these data, it is possible to compare different characteristics of radiosensitivity for the two cell lines in exponential and stationary (plateau) phases of growth in culture. Estimates of the parameters of model 5.1 obtained from the data by the method of least squares (with log S and log $\bar{S}$ in Formula 5.2 instead of S and $\bar{S}$) are given in Table 4. The corresponding estimates of the dose-effect curves are shown in Figures 2 and 3.

No significant difference between stationary and exponential states of growth was found for HeLa cells as far as numerical parameters of model 5.1 were concerned. It is not that surprising in view of the fact that, being irradiated in different phases of their growth, HeLa cells demonstrate virtually identical dose-effect patterns (Figure 2). On the contrary, marked differences manifest themselves for the two states of cell culture when results with 3T3 cells are examined (see Table 4 and Figure 3). In the stationary phase, 3T3 cells are considerably less sensitive than in the exponential phase as is reflected by the lower value of the parameter μ. At the same time, the critical number of radiation-induced lesions also appears to be lower for these cells in the stationary phase of their growth than in the exponential phase. This discrepancy can possibly be explained by a more efficient sublethal damage repair in the exponentially growing cells. Thus, the actively proliferating 3T3 cells simultaneously possess both a higher sensitivity to radiation (the mean number of the radiation-induced lesions per unit dose) and a greater capacity for sublethal damage repair (the critical number of hits) than their resting counterparts.

It is not a general regularity that the survival of cells irradiated in the exponential phase of cell culture growth is below that of cells exposed to radiation in the stationary phase. The typical example of the opposite situation is the survival of Chang liver cells (LICH) irradiated during different stages of their growth in cell culture.[96] In this case it is natural to expect that results of the data analysis will be drastically different from those obtained for 3T3 cells. Actually, as can be seen in Table 5, LICH cells appear to be twice as sensitive when they are irradiated in the stationary phase as compared with the exponential state of growth, but they have a higher capacity for

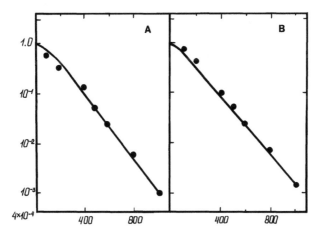

Figure 2
Dose-survival plot for plateau (A) and exponential (B) phases of growth for
HeLa S-3 cells. Solid lines = parametric estimates of the dose-effect curves.

repair of sublethal lesions. Using the survival data for LICH cells
subcultured not immediately but 6 h after irradiation in the stationary
phase, the contribution of potentially lethal damage repair may be
estimated.

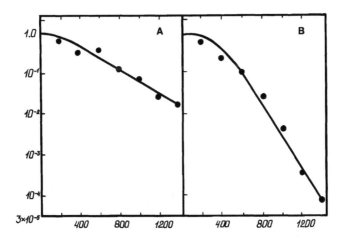

Figure 3
Dose-survival plot for plateau (A) and exponential (B) phases of growth for
3T3 cells. Solid lines = parametric estimates of the dose-effect curves.

Table 4

**Estimates of the Model Parameters
Resulting from the Analysis of
Experimental Data by Kim et al.[38]**

Cell Line	Growth Phase	$\hat{m}$	$\hat{\mu}$	$\hat{v}$
HeLa	Plateau	1	1.00	0.14
HeLa	Exponential	1	1.06	0.23
3T3	Plateau	4	1.24	0.38
3T3	Exponential	7	2.02	0.15

Note: Here, m is the critical number of hits; μ and $v = \sigma/\mu$ are the mean value and variance of the radiosensitivity distribution.

The results presented in Table 5 show that the value of μ is reduced if explantation of the cells for measuring their clonogenic capacity is delayed for 6 h, but the critical number of hits, m, remains the same as in the case of explantation immediately after the irradiation. These conclusions are in qualitative agreement with those drawn by Yakovlev and Zorin[90] as a result of their computer simulation study. For a more detailed biological discussion of this and other regularities in the survival of irradiated LICH cells, we refer the reader to the above-mentioned book by Yakovlev and Zorin.

The dose-survival plots presented in Figures 2, 3, 4, and 5 show that the randomized multihit-one target model given by Expression 5.1 provides a good description of irradiated cell survival. Parametric identification of the model is feasible and yields plausible interpretation of experimental evidence.

Table 5

**Estimates of the Model Parameters Resulting from
the Analysis of Experimental Data by Zinninger
and Little[97] for LICH Cells**

Experimental Conditions	$\hat{m}$	$\hat{\mu}$	$\hat{v}$
Exponential phase	2	0.99	0.18
Plateau phase (subculture immediately after irradiation)	4	1.88	0.25
Plateau phase (subculture 6 h after irradiation)	4	1.67	0.26

Note: Here, m is the critical number of hits; μ and $v = \sigma/\mu$ are the mean and variance of the radiosensitivity distribution.

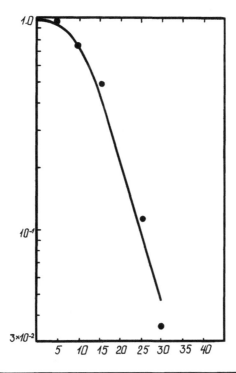

Figure 4
Cell-survival plot for LICH cells irradiated in the exponential phase of their growth. Theoretical curve (solid line) and experimental data by Zinninger and Little[96] are shown.

1.6   DISTRIBUTION OF CLONOGENS IN IRRADIATED TUMORS

Despite our intention to concentrate on theoretical aspects of the problem, we cannot avoid the following question: How could the probability of cell survival and the corresponding radiosensitivity distribution be estimated from available clinical data on the efficiency of cancer treatment? Certainly, there is always the possibility of inviting medical experts to provide the most plausible parameters of the model. But even a very experienced expert will be puzzled to give any recommendations unless some preliminary clinical studies yield quantitative information about the parameters of interest. At present, we are unable to provide the reader with ready-made methods for

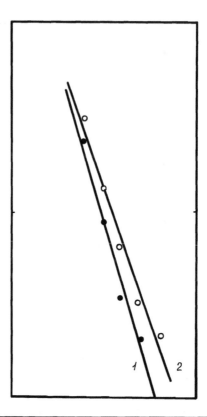

Figure 5
Cell-survival plots for LICH cells irradiated in the stationary (plateau) phase of growth. Subculture (1) immediately after irradiation and (2) 6 h after irradiation.

obtaining practical solution to the problem. What we can do is to outline briefly some principal ideas which seem to be promising and deserving of further development. Thus, the discussion presented in this section may be considered, in a sense, as a starting point of a more comprehensive biomathematical research.

We begin with a closely related topic, i.e., estimation of the mean number of those tumor cells that survive the treatment and retain capability of giving rise to tumor regeneration. We shall call such cells clonogenic or, following Tucker et al.,[85] clonogens. It is customary to use the terms "survival" and "clonogenic capacity" interchangeably as far as neoplastic cells are concerned. Consider the case when tumors are exposed to large single doses of radiation. In this case, it is natural to assume that the size N of the population of clonogens in a tumor just prior to irradiation is very large but the probability p of their survival after the treatment is very low. If N is nonrandom, one may confidently consider the number ν of surviving clonogens as a Poisson random variable. It is straightforward to write down the expression for the probability η of tumor cure (no surviving clonogens) in the form

$$\eta = \mathbb{P}\{\nu = 0\} = e^{-\Theta}, \quad \Theta = Np \quad , \tag{6.1}$$

where Θ is the mean number of clonogens surviving the treatment. A pertinent estimator for η is the observed proportion of patients that have been cured under given conditions, though there may be some complications caused by data censoring effects.[10,26,33] We shall touch upon this issue once again while discussing a more sound method that has been proposed for estimation of parameter Θ from the tumor recurrence data.

Suppose that at one's disposal are relevant dose-response data with regard to the frequency of tumor cure. Then, instead of 6.1, there arises the system of equations

$$\eta(D_i) = e^{-Np(D_i)}, \quad i = 1, \ldots, r \quad , \tag{6.2}$$

where D_i are doses of radiation for which information on dose-response relationship is available. In this situation, estimation of the survival probability for different dose values is feasible, at least in theory, and so is the underlying radiosensitivity distribution (see Section 1.5).

If N is random, then the situation is not that simple. There is one case, however, when Formula 6.2 retains its form, namely, when N has the Poisson distribution as well. Actually, in derivation of 6.1 or 6.2, we proceed from the description of cell survival within the framework of Bernoulli trials. Now the number of trials becomes random. As this number is assumed to be Poisson with parameter a, we may write

$$\mathbb{P}\{\nu = k\} = \sum_{n=k}^{\infty} \binom{n}{k} p^k(1 - p)^{n-k} \frac{a^n}{n!} e^{-a} =$$

$$\frac{(pa)^k}{k!} e^{-a} \sum_{m=0}^{\infty} \frac{(1 - p)^m a^m}{m!} = \frac{(pa)^k}{k!} e^{-pa} \quad .$$

Consequently, the form of Equations 6.1 and 6.2 remains unchanged, but N should be replaced by its expectation a.

In reality, this nice picture may happen to be seriously distorted. Usually, tumors are treated by a fractionated course of irradiation that allows cell proliferation to occur during the time intervals between successive fractions of radiation dose, and the distribution of clonogens surviving the final fraction is no longer expected to be Poisson. This does not exclude, however, a possibility that the two distributions are fairly close to each other. How well is the number of clonogens surviving such a regimen of radiotherapy described by a Poisson distribution? In this exact formulation, the problem was posed in the paper of Tucker et al.[85] We consider this paper to be very interesting and potentially important because until the time it was published, there had been no doubts that the Poisson statistics are appropriate for the description of the number of clonogens in treated tumors.[49,65,75,76] Therefore, it is worth discussing the results by Tucker et al. more thoroughly.

To highlight the problem formulated above, the authors conducted computer simulations. In order to briefly describe the design of their study, we introduce the following notations:

N = Initial number of clonogens (assumed to be nonrandom);

p = Probability of clonogen survival after one fraction of radiation dose;

n = Total number of dose fractions separated by equal time intervals;

α = Probability of cell division during the interval between two successive fractions.

Tucker et al. proceeded from the assumption that the probability p is the same for all clonogenic cells and for each fractional dose, and that it is also independent of the radiation dose given before. The same was assumed for the division probability α. In the simple version of the model used by Tucker et al., the asynchronous transition of cells to the proliferative state and temporal organization of the mitotic cycle were disregarded. In other words, cell proliferation activity was completely determined by the value of α. No significant discrepancies were observed when the results were compared with those obtained with the aid of a more complicated model involving an age-dependent probability of cell division. Therefore, we will dwell only upon the first part of this simulation study.

For a large number of tumors, each containing the same initial number N of clonogenic cells, the response to uniform regimens (i.e., those consisting of equal fractions) was simulated. A typical output of these simulations was the frequency of tumor cure as a function of the total dose given in equal fractions. In the first series of simulation experiments, studying the case $\alpha = 0$ (no cell proliferation during the treatment), the authors observed an excellent agreement with the predictions based on the Poisson distribution. We shall show that this result is, in fact, a general one and can be obtained easily without computer assistance.

In what follows, we shall use methods and terminology of the theory of branching stochastic processes. The reader who is not familiar with fundamentals of this theory is referred to Chapter 2 of the book by Jagers[32] or Chapter 1 of the book by Yakovlev and Yanev.[89] Let Z_t be the number of clonogens surviving t successive fractions of radiation dose. The random variable Z_t may be thought of as a Galton-Watson branching process with probability generating function $\varphi(s)$ for the number of offsprings in every generation t. The function $\varphi(s)$ is also called the reproduction generating function. If we proceed from the same assumptions as those used in the simulation model, the reproduction generating function for $t = 1, \ldots, n - 1$ is specified by the following expression:

$$\varphi(s) = 1 - p + p(1 - \alpha)s + p\alpha s^2, \quad |s| \leq 1 \quad . \tag{6.3}$$

Since we are not interested in cell proliferation that occurs after the nth fraction (but only in cell survival), for $t = n$, the reproduction generating function is equal to

$$\psi(s) = 1 - p + ps, \quad |s| \leq 1 \quad . \tag{6.4}$$

The quantities

$$h = \varphi'(1) \quad ,$$

$$\sigma = \varphi''(1) + h - h^2$$

are the reproduction mean and reproduction variance, respectively. From 6.3, we have

$$h = p(1 + \alpha) \quad ,$$

$$\sigma^2 = 2p\alpha + h - h^2 \quad . \tag{6.5}$$

These are individual probabilistic characteristics defined for every cell pertaining to the tth generation, $t = 1, \ldots, n - 1$. For $t = n$, the corresponding characteristics can be obtained from 6.4:

$$h_n = \psi'(1) = p \quad ,$$

$$\sigma_n = \psi''(1) + h_n - h_n^2 = p - p^2 \quad . \tag{6.6}$$

If the process starts from only one cell of the zeroth generation (one clonogen is present prior to the first irradiation), then the probability generating function for Z_t satisfies the recurrent relationship

$$\varphi_t(s) = \varphi_{t-1}(\varphi(s)) \quad ,$$

$$\qquad\qquad t = 1, \ldots, n - 1; \; |s| \le 1 \quad , \tag{6.7}$$

$$\varphi_0(s) \equiv s \quad ,$$

where $f(\varphi(\;))$ is the composition of functions.

Similarly, for the random variable Z_n, we have

$$\varphi_n(s) = \varphi_{n-1}(\psi(s)) \quad , \tag{6.8}$$

where φ_{n-1} is obtained from 6.7.

If the initial number of clonogens is equal to N, we have just to consider $[\varphi_t(s)]^N$ and $[\varphi_n(s)]^N$, where φ_t and φ_n are defined by 6.7 and 6.8, respectively. For random N, all formulas are generalized easily, but this is not that necessary in our discussion.

Consider the case $\alpha = 0$. In this case, there is no difference between φ and ψ, and we may write

$$\psi_t(s) = \psi_{t-1}(\psi(s)) \quad ,$$

$$\qquad\qquad t = 1, \ldots, n - 1; \; |s| \le 1 \quad , \tag{6.9}$$

$$\psi_0(s) \equiv s \quad ,$$

where $\psi(s)$ is given by 6.4.

After n iterations in 6.9, we get

$$\psi_n(s) = 1 - p^n + p^n s \quad .$$

Hence, for the probability generating function of the number of clonogens surviving the nth fraction of radiation dose, the following formula holds

$$[\psi_n(s)]^N = (1 - p^n + p^n s)^N$$

which corresponds to the binomial distribution with parameters N and p^n. If N is large, p^n is small, and $Np^n = \lambda$, then the Poisson theorem allows us to use the approximation

$$[\psi_n(s)]^N = e^{\lambda(s-1)} \quad ,$$

i.e., the distribution of clonogens surviving the nth irradiation is expected to be Poisson. Thus, it is not surprising that Tucker et al. obtained such good agreement between the predictions of the Poisson statistics and their numerical results concerning the probability of tumor cure (see 6.1 and 6.2) corresponding within the framework of branching processes to the extinction probability

$$\eta = [\psi_n(0)]^N = (1 - p^n)^N \quad .$$

When $\alpha > 0$, the distribution of the number of clonogens is no longer expected to be very close to the Poisson one. To elucidate this point, consider variances of the two distributions. Using recurrent Formula 6.7, it is not difficult to derive an expression for the variance of Z_t for $t = 1, \ldots, n - 1$. The resultant formula[89] is of the form

$$\mathcal{D}_t = \begin{cases} \dfrac{N\sigma^2 h^t(h^t - 1)}{h^2 - h}, & h \neq 1 \quad , \\[3mm] N\sigma^2 t, & h = 1 \quad , \end{cases} \qquad t = 1, \ldots, n - 1 \quad . \qquad (6.10)$$

We are interested primarily in the subcritical case, i.e., $h < 1$ or equivalently $p < \dfrac{1}{1 + \alpha}$. This is just the case that was simulated in the work of Tucker et al. Formula 6.8 allows us to obtain the variance $\mathcal{D}_n$, using the second derivative of the generating function at the point $s = 1$:

$$\varphi_n''(1) = \varphi_{n-1}''(1)[\psi'(1)]^2 + \varphi_{n-1}'(1)\psi''(1) \quad .$$

Consequently,

$$\mathcal{D}_n(t) = \varphi_n''(1) + M_n - M_n^2 =$$

$$(\mathcal{D}_{n-1} - M_{n-1} + M_{n-1}^2)h_n^2 + M_{n-1}(\sigma_n^2 - h_n + h_n^2) +$$

$$M_{n-1}h_n - M_{n-1}^2 h_n^2 = \mathcal{D}_{n-1}h_n^2 + M_{n-1}\sigma_n^2 \quad ,$$

where M_{n-1} is given by the formula

$$M_{n-1} = Np^{n-1}(1 + \alpha)^{n-1} = Nh^{n-1} \quad .$$

Taking 6.6 and 6.10 into account, we obtain finally

$$\mathcal{D}_n = Nh^{n-1}(p - p^2) + \frac{2Np^3\alpha h^{n-1}(1 - h^{n-1})}{h - h^2} +$$

$$Np^2h^{n-1}(1 - h^{n-1}) \quad ,$$

where $h = p(1 + \alpha)$.

For the mean number of clonogens surviving the nth fraction, the following formula holds:

$$M_n = h_n M_{n-1} = Nph^{n-1} = Np^n(1 + \alpha)^{n-1} \quad . \qquad (6.11)$$

The same formula should be used for the variance of Z_n, if we assumed the distribution of Z_n to be Poisson. Denoting this variance by V_n, we have by virtue of 6.11

$$V_n = Np^n(1 + \alpha)^{n-1} \quad .$$

To compare the values of $\mathcal{D}_n$ and V_n, consider the ratio

$$\frac{\mathcal{D}_n}{V_n} = 1 - p + \frac{2p^2\alpha(1 - h^{n-1})}{h(1 - h)} + p(1 - h^{n-1}), \; h < 1 \quad . \qquad (6.12)$$

The right-hand side of equality 6.12 is a monotonically increasing function of n and tends to a finite limit as $n \to \infty$. Consequently,

$$\frac{\mathcal{D}_n}{V_n} < 1 + \frac{2p^2\alpha}{h(1 - h)} \quad \text{for all } n \quad .$$

Substituting 6.5 for h in this inequality, we have the upper bound for the ratio $\mathcal{D}_n/V_n$ in the form

$$\frac{\mathcal{D}_n}{V_n} < 1 + \frac{2p\alpha}{(1 + \alpha) - p(1 + \alpha)^2} \quad . \qquad (6.13)$$

The right-hand side of 6.13 increases with α, increasing from 0 to $\min\left(\frac{1}{p} - 1, 1\right)$. Therefore, for $p < \frac{1}{2}$, the most unfavorable upper bound for the ratio $\mathcal{D}_n/V_n$ can be obtained by setting $\alpha = 1$ in 6.13, i.e.,

$$\frac{\mathcal{D}_n}{V_n} < 1 + \frac{p}{1 - 2p} \quad . \tag{6.14}$$

The claim that the value of p must be less than 0.5 ensures the sub-critical character of the process under consideration. Suppose, for example, that $p = 0.4$, then 6.14 yields the estimate $\mathcal{D}_n/V_n < 3$. The upper bound given by 6.14 decreases with decreasing p. If $p \to 0.5$ (the case $p = 0.5$, $\alpha = 1$ corresponds to the critical process), the right-hand side of 6.14 tends to infinity, thereby revealing considerable deviation of the perturbed variance $\mathcal{D}_n$ from V_n.

In simulations conducted by Tucker et al., the following values of the model parameters were used: $n = 30$, $p = 0.5$, and $\alpha = 0.4$. In this case $h = p(1 + \alpha) = 0.7$, and the corresponding branching process is subcritical. From 6.13 it follows that

$$\frac{\mathcal{D}_n}{V_n} < 1.95 \quad . \tag{6.15}$$

As n is large, the ratio $\mathcal{D}_n/V_n$ is very close to its upper bound given by 6.15. But this does not mean that $\mathcal{D}_n$ is much greater than V_n for every n. In particular, from 6.12, one might have obtained in the case at hand the following values of the ratio $\mathcal{D}_n/V_n$ for $n = 1, 2, 3$:

$$\frac{\mathcal{D}_1}{V_1} = 0.5, \quad \frac{\mathcal{D}_2}{V_2} = 0.94, \quad \frac{\mathcal{D}_3}{V_3} = 1.24 \quad .$$

Of course, in practical applications, we are most interested in possible discrepancies between $\mathcal{D}_n$ and V_n for large values of n. Therefore $\mathcal{D}_n/V_n = 1.95$ is a reasonable estimate in the example under consideration.

In reality, the discrepancy between $\mathcal{D}_n$ and V_n should be expected to be much less than that revealed by the above numerical example. The reasoning given by Tucker et al.[86] is worth quoting in this connection:

The largest amount by which Poisson statistics underestimated tumor cure in this study ($\approx$15%) was obtained when the tumor clonogen doubling time was assumed to be 2.06 days throughout the course of a 30-fraction treatment given as one fraction per day; in terms of total dose, this corresponded to a discrepancy of about 2–3 Gy. Although many tumors may have clonogen doubling times as short as 2 days by the end of treatment, it is unlikely that proliferation would be that rapid, in general, until the tumor had been exposed to a significant proportion of the total dose. Based on these results, we estimate that the error in Poisson statistics in describing tumor cure is unlikely to be much more than about 10% after most standard treatment regimens, and that it would often be much less than that. However, the error could be greater for certain nonstandard regimens, e.g., those with an unusually long break near the end of treatment.

In their simulation experiments, Tucker et al. also tested estimation of unknown parameters based on the maximum likelihood method. In particular, they estimated the basic parameters by fitting a model of the dose-response type to tumor-cure data which were simulated under the assumption that there was a break in treatment of 5, 10, or 15 d occurring after 2 weeks of the daily irradiation. There was a very good agreement between estimated and actual values of a parameter related to the survival probability p, but the estimate of ln N was found to be biased downward. In this connection, it should be especially noted that it is the value of p we are concerned with in the optimization problems representing the main subject of Chapters 2 and 3 of this book.

We share the authors' opinion that the value of division probability $\alpha = 0.4$ is unrealistically high in view of typical durations of the mitotic cycle in tumors and asynchronous entry of cells into the prereplicative period after irradiation. Furthermore, we would like to stress one more factor that might substantially decrease the expected value of α, i.e., radiation-induced block of DNA synthesis and mitosis. It is a well-known fact that the mean duration of the radiation block frequently exceeds a 1-d interval chosen in the simulation study by Tucker et al. (see discussion of this issue in Reference 90). We do believe that the Poisson distribution will keep playing a key role in parametric analysis of the efficiency of cancer treatment. There is another problem with this distribution that arises when the probability of tumor cure given by 6.1 is used for estimation of the mean number of clonogens in irradiated tumors. The probability η represents a tail defect of the survival function defined for the life length of a patient treated for cancer. Due to censoring effects inherent to data on survival of oncological patients,[10,26,33] the frequency of tumor cure determined for a specified period of observation, e.g., 5 years, cannot serve as an appropriate estimator of the probability η. Even the Kaplan-Meier estimator (nonparametric estimator adjusted to analysis of censored observations) is not good for this purpose because it appears to be highly unstable in the region of tail defect in the presence of heavy censoring.[62] It is reasonable, therefore, to introduce time factor in the parametric model aimed at estimating the mean number of clonogens in irradiated tumors. In other words, it is desirable to find a link (and to formalize it mathematically) between the parameter Θ (see Formula 6.1) and survival of oncological patients for the whole period of observation. This cannot fail to supply an estimation procedure with additional information about unknown parameters of the model, the mean number of clonogens included. Such a method will be proposed below.

To illustrate the principal idea, consider the case of true recurrence caused by clonogens that had survived the treatment. Each surviving clonogenic cell possesses, in the long run, the capacity for giving rise to an overt tumor. Let X_i be a random time for the ith cell to produce a detectable tumor. By analogy with the terminology accepted in carcinogenesis studies, we call X_i a progression time. Nonnegative random variables X_i, $i = 1, 2, \ldots$ are assumed to be independent and identically distributed with the common distribution function $F(x)$. The time to the tumor recurrence can be defined as

$$U = \min_{0 \le i \le \nu} X_i \quad ,$$

where ν is the number of clonogens surviving the treatment. If ν is a Poisson random variable independent of the sequence $X_1, X_2, \ldots,$ the survival function $G(t)$ for the random variable U can be obtained easily as follows:

$$G(t) = \mathbb{P}\{U \ge t\} = \sum_{k=0}^{\infty} \frac{\Theta^k}{k!} e^{-\Theta}[1 - F(t)]^k = e^{-\Theta F(t)} \quad . \quad (6.16)$$

In this expression, two substantive characteristics of tumor growth are interwoven: the mean number of surviving clonogens, Θ, and the rate of their progression described by the function F. The parameter Θ, as well as numerical parameters of the distribution function, F, can be estimated from time-to-recurrence observations. The survival function G corresponds to improper (substochastic) distribution, and its limit value $G(+\infty) = e^{-\Theta}$ represents the probability of tumor cure (compare with Formula 6.1).

The hazard function $\lambda(t)$, defined with respect to $G(t)$, is given by the equality

$$\lambda(t) = \Theta f(t) \quad ,$$

where f is the density of the distribution F. If the progression time distribution F is unimodal, then the hazard function $\lambda(t)$ has a maximum. Note that the assumption on the exponential form of F, $F(t) = 1 - e^{-at}$, $t \geq 0$, should be rejected at once because in this case, the hazard function λ is monotonically decreasing in time, which is unrealistic.

To describe a possible heterogeneity of clonogens with respect to the progression time distribution, introduce k different types of tumor cells with distributions $F_j(t)$. Then the progression time distribution F is represented by a finite mixture:

$$F(t) = \sum_{j=1}^{k} q_j F_j(t), \ 0 < q_j < 1, \ \sum_{j=1}^{k} q_j = 1 \quad .$$

This mixture of distributions yields the independent competing risks model for the function G, i.e.,

$$G(t) = \prod_{j=1}^{k} \exp \{-\Theta_0 q_j F_j(t)\} \quad , \tag{6.17}$$

where Θ_0 is the expected total number of viable clonogens of various types existing in the irradiated tumor. Within the framework of this model, the hazard functions λ_j are additive and

$$\lambda(t) = \Theta_0 \sum_{j=1}^{k} q_j f_j(t) \quad .$$

In view of the last formula, it is not surprising that the bimodal shape of hazard function arises when tumor recurrences originate from two distinct subpopulations of progenitor cells as in the example presented below.

We have applied the distribution given by 6.16 to analysis of data on cancer of cervix uteri. This analysis was accomplished in collaboration with Professor J. Trelford (University of California, Davis). Two samples of the time-to-recurrence data were under consideration: those referred to as the exophytic and the endophytic type of carci-

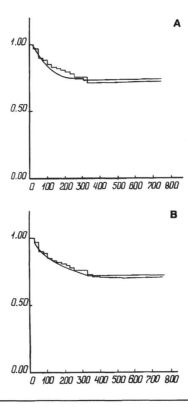

Figure 6
Estimation of the survival function G(t) for exophytic tumors. Stepwise curve
= Kaplan-Meier estimate; solid line = parametric estimate. (A) one fraction
of clonogens; (B) two fractions of clonogens.

noma of the cervix. For each of the samples, the Kaplan-Meier estimates of the survival functions G(t) were constructed (Figures 6 and
7). These nonparametric estimates differ significantly from each other.
The statistical hypothesis of homogeneity tested for the two groups
of patients was rejected even when three different statistical tests[13,15,48]
were used. The parametric analysis was based on Formula 6.16. If
one selects, in accordance with the principle of parsimony, a two-
parameter family of distributions to approximate the function F in
6.16, then there will be only three parameters to be estimated from
the time-to-recurrence observations, the estimation of which is feasible from available data, using the maximum likelihood method. In
selecting an appropriate functional form of F, the two-parameter gamma
distribution is a natural choice by virtue of its flexibility and clear

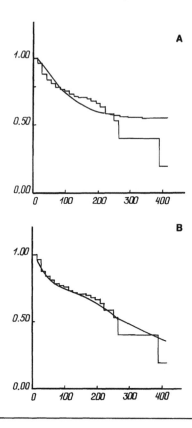

Figure 7
Estimation of the survival function G(t) for endophytic tumors. Stepwise curve = Kaplan-Meier estimate. (A) one fraction of clonogens; (B) two fractions of clonogens.

meaning of shape and scale parameters, hereafter denoted by a and b, respectively. These parameters are related to the mean $\bar{\tau}$ and the variance σ of the progression time as follows: $\bar{\tau} = a/b$, $\bar{\sigma}^2 = a/b^2$. To find their statistical estimates, the maximum likelihood method modified for the scheme of independent right-hand censoring[33] was used in analysis of time-to-recurrence data. The search for the maximum of likelihood function was carried out numerically by computer, using the algorithm of random search[94] followed by local optimization.[16]

Shown in Figures 6 and 7 are the Kaplan-Meier estimates and the corresponding parametric estimates of the survival functions for exophytic (Figure 6A) and endophytic (Figure 7A) tumors of the cervix uteri. A rather poor agreement between the two estimates (nonparametric and parametric) is evident; therefore one might assume a more complicated structure of tumor cell population in both cases.

Proceeding from the hypothesis about the presence in treated tumors of two distinct subpopulations of clonogens differing in their progression time distributions, it is possible to obtain the MLE of a larger number of the model parameters. For exophytic tumors, these estimates are $\Theta_1 = 0.16$, $\bar{\tau}_1 = 49$, $\sigma_1 = 35$, for the first fraction of clonogens, and $\Theta_2 = 0.20$, $\bar{\tau}_2 = 249$, $\sigma_2 = 124$, for the second one. The results for endophytic tumors are as follows: $\Theta_1 = 0.29$, $\bar{\tau}_1 = 42$, $\sigma_1 = 30$, for the first fraction of clonogens, and $\Theta_2 = 3.00$, $\bar{\tau}_2 = 566$, $\sigma_2 = 283$, for the second one.

In view of Formula 6.17, parameters Θ_j, $j = 1, 2$, are defined as $\Theta_j = \Theta_0 q_j$. The corresponding parametric estimates of the survival functions are depicted in Figure 6B and Figure 7B. These estimates provide a good description of the data as it can be seen in Figure 6B for exophytic and in Figure 7B for endophytic tumors. Hence, we conclude that the population of clonogens surviving treatment in both cases consists of two cell subpopulations which are drastically different with respect to the rate of developing tumor recurrence. In both types of tumor, the rapid fractions possess similar temporal characteristics, while there is a great discrepancy between the mean values of the progression time for exophytic and endophytic tumors. The above example is only an illustration of the proposed approach. Further analysis of real data, side by side with computer simulation studies, might have provided a more reliable substantiation of this parametric method as applied to tumor recurrence phenomena. Associated statistical problems are awaiting their solution as well.

1.7 INCOMPLETE REPAIR MODELS OF CELL SURVIVAL FOR FRACTIONATED AND CONTINUOUS IRRADIATION REGIMENS

In our exposition of incomplete repair (IR) models, we follow the approach of Thames[81] and amplify it by some generalizations.

The IR model is derived under the assumptions that radiosensitivity of cells remains constant in the course of irradiation and that cell proliferation is negligible. The key idea is to describe the cumulative effect of radiation damage and its subsequent repair by the "dose-equivalent of incomplete repair".[54]

It is convenient to characterize cell survival response to a single (acute) dose D by the value $g(D) = -\ln S(D)$, where $S(D)$ is the

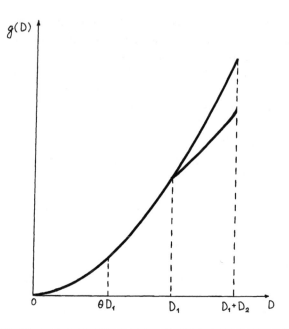

Figure 8
See text for explanations.

survival probability (or surviving fraction) (see upper curve in Figure 8). Obviously, $g(0) = 0$.

To describe survival of a cell exposed to two doses, D_1 and D_2, separated by interval τ, it is envisioned that the dose-survival curve for two-fraction exposure is repeating, starting from the point D_1, a segment of the original survival curve beginning at some point ΘD_1 with $\Theta = \Theta(\tau)$. Hence (Figure 8),

$$g(D_1, D_2; \tau) = g(D_1) + g(\Theta D_1 + D_2) - g(\Theta D_1) \quad . \quad (7.1)$$

The value ΘD is called the dose-equivalent of incomplete repair for the time τ elapsed from irradiation with dose D.

In the case $\tau = 0$, i.e., in that of single exposure to the dose $D_1 + D_2$, $g(D_1, D_2, \Theta)$ should lie on the original survival curve, which means that $\Theta(0) = 1$. Accordingly, Formula 7.1 yields $g(D_1, D_2, 0) = g(D_1 + D_2)$. In another extreme case of complete repair ($\tau \to +\infty$), the whole initial shoulder of the original curve is repeated. Therefore, $\Theta(+\infty) = 0$, and 7.1 converts into

$$g(D_1, D_2; +\infty) = g(D_1) + g(D_2) \quad ,$$

the usual formula for two-fraction survival in the case of complete repair.

According to Oliver's model,[81]

$$\Theta(\tau) = e^{-\mu\tau}, \quad \mu > 0 \quad, \tag{7.2}$$

which is consistent with the above-mentioned extreme cases.

For a triple of doses D_1, D_2, D_3, separated by two intervals τ_1 and τ_2, the equivalent of incomplete repair related to the dose D_1 equals $\Theta_1\Theta_2 D_1$ and to the dose D_2, $\Theta_2 D_2$. Hence,

$$g(D_1, D_2, D_3; \tau_1, \tau_2) = g(D_1) + g(\Theta_1 D_1 + D_2) - g(\Theta_1 D_1) +$$

$$g(\Theta_1\Theta_2 D_1 + \Theta_2 D_2 + D_3) - g(\Theta_1\Theta_2 D_1 + \Theta_2 D_2) \quad .$$

Continuing this way of thinking, we obtain for an arbitrary multiple dose regimen $(\vec{D}; \vec{\tau})$ where $\vec{D} = (D_1, \ldots, D_n)$ is a sequence of doses and $\vec{\tau} = (\tau_1, \ldots, \tau_{n-1})$, is that of interfraction intervals,

$$g(\vec{D}; \vec{\tau}) = \sum_{k=1}^{n} \left[g\left(\sum_{j=1}^{k} D_j \prod_{i=j}^{k-1} \Theta_i \right) - g\left(\sum_{j=1}^{k-1} D_j \prod_{i=j}^{k-1} \Theta_i \right) \right] \tag{7.3}$$

(note that by definition the sum and the product over the empty set of indices is set to be 0 and 1, respectively). For dose-equivalents of IR of the form 7.2, we have

$$g(\vec{D}; \vec{\tau}) =$$

$$\sum_{k=1}^{n} \left[g\left(\sum_{j=1}^{k} D_j \exp\left(-\mu \sum_{i=j}^{k-1} \mu_i \right) \right) - g\left(\sum_{j=1}^{k-1} D_j \exp\left(-\mu \sum_{i=j}^{k-1} \tau_i \right) \right) \right] \quad .$$

In the particular case of n equals doses D separated by equal intervals of length τ, 7.3 acquires the form

$$g_n(D; \tau) = \sum_{k=1}^{n} \left[g\left(D \sum_{k=0}^{n-1} \Theta^k \right) - g\left(D \sum_{k=1}^{n-1} \Theta^k \right) \right]$$

(see Formula 5 in Reference 81).

In the case $\tau_i = 0$ $(\Theta_i = 1)$, $i = 1, \ldots, n - 1$, we get from 7.3

$$g(\vec{D};0) = g\left(\sum_{k=1}^{n} D_k \right), \text{ i.e., response to the total dose of the size}$$

$\sum_{k=1}^{n} D_k$, while in the opposite case of large τ_i ($\Theta_i = 0$), $i = 1, \ldots,$ $n - 1$, referring to complete repair, formula 7.3 implies

$$g(\vec{D}; +\infty) = \sum_{k=1}^{n} g(D_k) \quad .$$

To proceed with the IR model, we need to choose a survival function S(D).

In the small dose per fraction limit, one can use the linear-quadratic (LQ) approximation

$$S(D) = e^{-(\alpha D + \beta D^2)}, \quad \alpha, \beta > 0 \quad ,$$

or equivalently

$$g(D) = \alpha D + \beta D^2 \quad .$$

For such g, when calculation in 7.3 is carried out, the result is

$$g(\vec{D}; \vec{\tau}) = \sum_{k=1}^{n} g(D_k) + 2\beta \sum_{k=1}^{n} \sum_{j=1}^{k-1} D_k D_j \prod_{i=j}^{k-1} \Theta_i \quad . \tag{7.4}$$

For equal doses of the size D and equal intervals between them, this is

$$g_n(D; \tau) = ng(D) + 2\beta D^2 \frac{\Theta}{1 - \Theta} \left(n - \frac{1 - \Theta^n}{1 - \Theta} \right) \tag{7.5}$$

(compare with Formula 7 in Reference 81).

As it follows from 7.4, in the LQ model, IR affects only the quadratic component of lesion formation.

In the particular case $\beta = 0$, the LQ model reduces to the multihit-one target model with $m = 0$. In this model $g(D) = xD$, where $x = \alpha$ can be interpreted as the radiosensitivity of a cell. In view of 7.4,

$$g(\vec{D}; \vec{\tau}) = \sum_{k=1}^{n} g(D_k) = x \sum_{k=1}^{n} D_k \quad .$$

In other words, survival response to a regimen $(\vec{D}; \vec{\tau})$ of irradiation is equal to that for acute exposure to the dose $\sum\limits_{k=1}^{n} D_k$. In the more general multihit model,

$$S(D) = e^{-xD}\left(1 + xD + \ldots + \frac{(xD)^m}{m!}\right) \quad .$$

Therefore,

$$g(D) = xD - \ln\left(1 + xD + \ldots + \frac{(xD)^m}{m!}\right) \quad .$$

Hence,

$$g(\vec{D}, \vec{\tau}) = x\sum\limits_{k=1}^{n} D_k - h(\vec{D}; \vec{\tau}) \quad ,$$

where $h(D) = \ln\left(1 + xD + \ldots + \frac{(xD)^m}{m!}\right)$. Expression 7.3 for such function h, though looking cumbersome, can be calculated explicitly. This allows for solution of various optimization problems in multifractional setting within the multihit-one target model (see Chapter 3).

The IR model can also be applied to continuous exposures with constant dose rate. Suppose the total dose D is administered during time T at the constant dose rate $v = D/T$. Then we may define the response function by the formula

$$g(D; T) = \lim\limits_{n\to\infty} g_n\left(\frac{D}{n}; \frac{T}{n-1}\right) \quad .$$

In the framework of the LQ model, we apply 7.5 to obtain

$$g(D; T) = g(D) + \beta D^2\varphi(\mu T) \tag{7.6}$$

with

$$\varphi(t) = \frac{2}{t^2}\left(e^{-t} - 1 + t - \frac{t^2}{2}\right)$$

(see Formula 8 in Reference 81). Expression 7.6 also appears in the Roesch accumulation model.[69]

The IR model, both in its discrete and continuous modifications, was tested in numerous experiments and proven to provide reasonably accurate predictions of cell survival (see Reference 81 and references therein). Nevertheless, we will compare the results of the IR model with those of the more biologically grounded lethal, potentially lethal (LPL) model,[12] involving repair-misrepair considerations[84] and the cybernetic model of Pohlit and Heyder.[64]

The model in question presupposes division of lesions into repairable and irreparable ones. It is assumed that the expected amount of primary lesions of each type induced directly by irradiation is proportional to the dose. The rate of repair process is proportional to the number of repairable lesions. The process of (pairwise) interaction and fixation of lesions producing irreparable lesions proceeds at a rate proportional to the square of the number of unrepaired primary lesions. Fixation of lesions can be also interpreted as "misrepair".[84]

A single (acute) exposure to the dose D induces the expected numbers x_1D and x_2D of repairable and irreparable lesions, respectively. Denote by $N_1(t)$ and $N_2(t)$ their expected amounts at the moment t. The evolution of lesions is described by the following system of ordinary differential equations:

$$\frac{dN_1}{dt} = -\epsilon_1 N_1 - \epsilon_2 N_1^2, \quad N_1(0) = x_1D \quad ,$$

$$\frac{dN_2}{dt} = \epsilon_2 N_1^2, \qquad\qquad N_2(0) = x_2D \quad , \qquad (7.7)$$

where ϵ_1, ϵ_2 are positive constants. Integrating these equations, we get readily

$$N_1(t) = \frac{x_1 D \Theta}{1 + \frac{x_1 D}{\epsilon}(1 - \Theta)} \quad ,$$

$$N_2(t) = (x_1 + x_2)D - \frac{x_1 D \Theta}{1 + \frac{x_1 D}{\epsilon}(1 - \Theta)} -$$

$$\epsilon \ln\left(1 + \frac{x_1 D}{\epsilon}(1 - \Theta)\right) \quad , \qquad (7.8)$$

where $\Theta = e^{-\epsilon_1 t}$ and $\epsilon = \epsilon_1/\epsilon_2$. We conclude from 7.8 that $N_1(t) \to 0$ as $t \to +\infty$ and that

$$N_1 + N_2 = xD - \epsilon \ln\left(1 + \frac{x_1 D}{\epsilon}(1 - \Theta)\right)$$

with $x = x_1 + x_2$.

We now identify the survival function $S(D)$ with the probability of presence of at least one irreparable lesion. Since the process of their formation can be viewed as the Poisson one, we find that

$$S(D) = \lim_{t \to +\infty} e^{-N_2(t)} = \lim_{t \to +\infty} e^{-(N_1(t) + N_2(t))} = e^{-xD + \epsilon \ln\left(1 + \frac{x_1 D}{\epsilon}\right)} =$$

$$e^{-xD}\left(1 + \frac{x_1 D}{\epsilon}\right)^\epsilon .$$

Hence,

$$g(D) = xD - \epsilon \ln\left(1 + \frac{x_1 D}{\epsilon}\right)$$

(note that in the particular case $x_2 = 0$, the above expression for $S(D)$ coincides with 3.9). For small doses D, we obtain by expanding the logarithm

$$g(D) \sim x_2 D + \frac{x_1^2}{2\epsilon} D^2, \quad D \to 0 ,$$

which brings us back to the LQ model with $\alpha = x_2$ and $\beta = \frac{x_1^2}{2\epsilon}$.

Following the idea of Thames,[81] it is convenient to write the solution of system 7.7 using evolution operators. If $N_1(t)$, $N_2(t)$ are initial conditions at a moment t, then in the absence of irradiation during the period $(t, t + \tau)$, we have

$$N_1(t + \tau) = U_\tau(N_1(t)) ,$$

$$N_2(t + \tau) = N_2(t) + V_\tau(N_1(t)) , \qquad (7.9)$$

where

$$U_\tau(s) = \frac{s\Theta}{1 + \frac{s}{\epsilon}(1 - \Theta)} ,$$

$$V_\tau(s) = s - \frac{s\Theta}{1 + \frac{s}{\epsilon}(1 - \Theta)} - \epsilon\ln\left(1 + \frac{s}{\epsilon}(1 - \Theta)\right)$$

with $\Theta = e^{-\epsilon_1\tau}$. Employing Taylor expansions, we see easily that for $s \to 0$,

$$U_\tau(s) = s\Theta + O(s^2)$$

$$V_\tau(s) = \frac{s^2}{2\epsilon}(1 - \Theta^2) + O(s^3) \quad . \tag{7.10}$$

Let t_k be the moments of exposure to doses D_k, $k = 1, \ldots, n$, $t_1 = 0$, $t_k = \sum_{i=1}^{k-1} \tau_i$, $k = 2, \ldots, n$. In view of 7.9, we have the following recurrence of initial conditions:

$$N_1(t_k) = x_1D_k + U_{\tau_{k-1}}(N_1(t_{k-1})) \quad ,$$

$$N_2(t_k) = x_2D_k + N_2(t_{k-1}) + V_{\tau_{k-1}}(N_1(t_{k-1})) \quad . \tag{7.11}$$

Now we iterate 7.11 to obtain for $t > t_n$:

$$N_2(t) = x_2 \sum_{k=1}^{n} D_k + V_{\tau_1}(x_1D_1) + V_{\tau_2}(x_1D_2 + U_{\tau_1}(x_1D_1)) +$$

$$V_{\tau_3}(x_1D_3 + U_{\tau_2}(x_1D_2 + U_{\tau_1}(x_1D_1))) + \ldots +$$

$$V_{\tau_{n-1}}(x_1D_{n-1} + U_{\tau_{n-2}}(x_1D_{n-2} + U_{\tau_{n-3}}(x_1D_{n-3} + \ldots) \ldots)) +$$

$$V_{t-t_n}(x_1D_n + U_{\tau_{n-1}}(x_1D_{n-1} + U_{\tau_{n-2}}(x_1D_{n-2} + \ldots) \ldots)) \quad .$$

Letting $t \to +\infty$, we get on account of 7.10 for small doses D_k, $k = 1, \ldots, n$,

$$N_2(+\infty) \sim x_2 \sum_{k=1}^{n} D_k + \frac{x_1^2D_1^2}{2\epsilon}(1 - \Theta_1^2) + \frac{x_1^2(D_2 + \Theta_1D_1)^2}{2\epsilon}(1 - \Theta_2^2) +$$

$$\frac{x_1^2(D_3 + \Theta_2(D_2 + \Theta_1D_1))^2}{2\epsilon}(1 - \Theta_3^2) + \ldots +$$

$$\frac{x_1^2}{2\epsilon}(D_{n-1} + \Theta_{n-2}(D_{n-2} + \ldots) \ldots)^2(1 - \Theta_{n-1}^2) +$$

$$\frac{x_1^2}{2\epsilon} (D_n + \Theta_{n-1}(D_{n-1} + \ldots))^2 =$$

$$x_2 \sum_{k=1}^{n} D_k + \frac{x_1^2}{2\epsilon} [D_1^2(1 - \Theta_1^2) + (D_2 + \Theta_1 D_1)^2(1 - \Theta_2^2) +$$

$$(D_3 + \Theta_2 D_2 + \Theta_2 \Theta_1 D_1)^2(1 - \Theta_3^2) + \ldots +$$

$$(D_{n-1} + \Theta_{n-2} D_{n-2} + \ldots + \Theta_{n-2} \Theta_{n-3} \ldots \Theta_1 D_1)^2(1 - \Theta_{n-1}^2) +$$

$$(D_n + \Theta_{n-1} D_{n-1} + \ldots + \Theta_{n-1} \Theta_{n-2} \ldots \Theta_1 D_1)^2] = x_2 \sum_{k=1}^{n} D_k + \frac{x_1^2}{2\epsilon} f(\vec{D}; \vec{\Theta}) \ .$$

The following identity, generalizing Formula 31 in Reference 81, can be easily checked by induction:

$$f(\vec{D}; \vec{\Theta}) = \sum_{k=1}^{n} D_k^2 + 2 \sum_{k=1}^{n} \sum_{j=1}^{k-1} D_k D_j \prod_{i=j}^{k-1} \Theta_i \ . \qquad (7.12)$$

As mentioned above, we suppose that even one irreparable lesion is lethal, therefore $g(\vec{D}; \vec{\tau}) = N_2(+\infty)$. Invoking our recent expression for $N_2(+\infty)$ and comparing 7.12 with 7.4, we conclude that, in the small dose per fraction limit, the LPL model for fractionated irradiation coincides with the IR model based on the LQ response function with coefficients α and β identified in the same way as for the case of single exposure, i.e., $\alpha = x_2$, $\beta = x_1^2/2\epsilon$.

We shall now formulate an analog of LPL model 7.7 related to continuous exposures and show that for small dose rates, we regain the survival function 7.6.

The system of differential equations describing repair-misrepair kinetics is modified to

$$\frac{dN_1}{dt} = x_1 v - \epsilon_1 N_1 - \epsilon_2 N_1^2, \quad N_1(0) = 0 \ ,$$

$$\frac{dN_2}{dt} = x_2 v + \epsilon_2 N_1^2, \qquad N_2(0) = 0 \ . \qquad (7.13)$$

The solution of 7.13 is

$$N_1(t) = \frac{2x_1 v(1 - \gamma)}{(1 - \gamma)\epsilon_1 + (1 + \gamma)\delta}, \quad N_2(t) = x_2 D + \epsilon_2 \int_0^t N_1^2(s) ds \ ,$$

where $\gamma = e^{-t\delta}$ and $\delta = \sqrt{\epsilon_1^2 + 4\epsilon_2 x_1 v}$. For small v, we have $\delta \approx \epsilon_1$.

Therefore,

$$N_1(t) \approx \frac{x_1 v}{\epsilon_1} (1 - \Theta),$$

$$N_2(t) \approx x_2 D + \frac{(x_1 v)^2}{2\epsilon\epsilon_1^2} (-2 \ln\Theta - 3 + 4\Theta - \Theta^2)$$

with $\Theta = \epsilon^{-\epsilon_1 t}$ as above and $D = vT$ being the total dose absorbed during the irradiation period T. To determine the response function g, we find an acute exposure to certain doses resulting in the same numbers of lesions $N_1(T)$, $N_2(T)$ and define g to be the value $N_2(+\infty)$ for the corresponding solution of 7.7. Thus, as we have already seen

$$g = N_2(t) + V_\infty(N_1(T)) \quad .$$

Therefore, for small enough v,

$$g(D; T) \approx x_2 D + \frac{(x_1 v)^2}{2\epsilon\epsilon_1^2} (-2 \ln\Theta - 3 + 4\Theta - \Theta^2) + \frac{(x_1 v)^2}{2\epsilon\epsilon_1^2} (1 - \Theta)^2 =$$

$$x_2 D + \frac{(x_1 v)^2}{\epsilon\epsilon_1^2} (-\ln\Theta + \Theta - 1) = x_2 D + \frac{x_1^2}{2\epsilon} D^2 + \frac{x_1^2}{2\epsilon} D^2 \varphi(\epsilon_1 T) \quad ,$$

which is just 7.6 for our conventional identifications $\alpha = x_2$, $\beta = x_2/2\epsilon$, and $\mu = \epsilon_1$.

2

Theoretical Bounds for the Efficiency of Cancer Treatment

2.1 INTRODUCTION

Extensive experimental studies[5] have shown that a considerable number of different factors manifesting themselves at the cellular level are responsible for radiation effects. Among them are spatial heterogeneity of tumor cell population and variation of cellular radiosensitivity depending on the position of a cell in its life cycle. It seems very difficult to take all the acting factors into account within the framework of a mathematical or even a simulation model.[39,80,90] Here, one faces the common dilemma: to choose a simple model which may fall far short of adequate description of reality or a complex one which needs to be supplied with vast experimental information and frequently renders optimization problems computationally unfeasible.

In particular, this is valid for the mathematical description of heterogeneity of cells with respect to their radiosensitivity. The conventional approach presupposes distinguishing within tumor tissue radioresistant cell reserves (hypoxic cells in solid tumors) which should have been exhausted as a result of reoxygenation and cell multiplication processes (for example, see Reference 77). A more natural way of introducing cell population heterogeneity has been presented by Rachev and Yakovlev.[68] They proposed a new criterion of cancer treatment efficiency, i.e., the difference between expected survival probabilities for normal and neoplastic cells, the expectation being taken with respect to the distributions of cell radiosensitivity.

Thus, the efficiency functional assumes the form

$$\Phi(\mathcal{D}) = \int_0^\infty g(x, \mathcal{D})d\mu_0(x) - \int_0^\infty g(x, \mathcal{D})d\mu_1(x) =$$

$$\int_0^\infty g(x, \mathcal{D})d(\mu_0 - \mu_1)(x) \quad , \tag{1.1}$$

where $\mathcal{D} = (D_1, \ldots, D_n)$, $D_i \geq 0$, $i = 1, \ldots, n$, is a sequence of irradiation doses, x is the radiosensitivity of a cell with respect to a given radiation, $g(x, \mathcal{D})$ is the survival probability of a cell with radiosensitivity x exposed to the sequence of doses $\mathcal{D}$, and μ_0, μ_1 are radiosensitivity distributions (viewed as probabilistic measures on $\mathbb{R}_+$) for normal and neoplastic cells, respectively.

The survival probability $g(x, \mathcal{D})$ as a function of $n + 1$ variables has the following natural properties: (1) it is nonincreasing with respect to each argument — the others being fixed; (2) $g(0, \mathcal{D}) = g(x, \mathbb{O}) = 1$ for all $x \in \mathbb{R}_+$ and $\mathcal{D} \in \mathbb{R}_+^n$; (3) $g(x, \mathcal{D}) \to 0$ as one of the arguments tends to infinity and the others are fixed.

Some principal optimization problems based on the functional 1.1 are considered in the subsequent sections.

2.2   BASIC ASSUMPTIONS AND PROBLEMS

The general problem in question is to find the value

$$\Phi^*(\Delta): = \sup_{\mathcal{D} \in \Delta} \Phi(\mathcal{D}) \tag{2.1}$$

and the corresponding optimal irradiation schemes $\mathcal{D}^*$ provided that the value $\Phi^*(\Delta)$ is attained by the functional Φ on the given set of schemes Δ. The following sets Δ seem to be biologically meaningful:

- $\Delta_0: = \{\mathcal{D} = (D_1, \ldots, D_n): D_i \geq 0, i = 1, \ldots, n, n \in \mathbb{N}\}$ (set of all irradiation schemes);
- $\Delta(D): = \{\mathcal{D} \in \Delta_0: D_1 + \ldots + D_n = D\}$, $D > 0$, (set of all fractionations of a given total dose D);
- $\tilde{\Delta}(D): = \{\mathcal{D} \in \Delta_0: D_1 + \ldots + D_n \leq D\}$, $D > 0$, (set of irradiation schemes with the total dose not greater than D);
- $\Delta^\gamma: = \{\mathcal{D} \in \Delta_0: \int_0^\infty g(x, \mathcal{D})d\mu_0(x) \geq \gamma\}$, $0 < \gamma < 1$, (set of such

irradiation schemes that guarantee the prescribed level γ of the expected survival for normal cells);

- $\Delta_N: = \{\mathcal{D} \in \Delta_0 : n \leq N\}$, $N \in \mathbb{N}$, (set of schemes with the bounded number of fractions);
- $\Delta_N(D): = \Delta(D) \cap \Delta_N$, $\tilde{\Delta}_N(D): = \tilde{\Delta}(D) \cap \Delta_N$, $\Delta_N^{\gamma}: = \Delta^{\gamma} \cap \Delta_N$ (appropriate intersections of the above mentioned sets).

To specify the functional Φ, we proceed from the following assumptions.

2.2.1 Assumption 1

Time intervals elapsing between the doses $D_1, \ldots, D_n$ are long enough for accomplishment of the transient processes of inactivation and recovery of damaged cells. This implies the representation

$$g(x, \mathcal{D}) = \prod_{i=1}^{n} g(x, D_i) \quad . \tag{2.2}$$

Thus, the survival probability does not depend on the order of doses constituting a given irradiation scheme.

2.2.2 Assumption 2

Normal and neoplastic cells differ only in their radiosensitivity distributions. It is the survival (clonogenic capacity) of neoplastic and normal cells that completely determines the ultimate effect of treatment. The radiosensitivity of a cell is inherited by its progeny without any changes. Thus, the radiosensitivity distributions μ_0 and μ_1 are not influenced by cell proliferation processes.

2.2.3 Assumption 3

The influence of tumor growth during radiotherapy upon the efficiency of treatment may be considered as negligible.

To describe the effect of irradiation in the case of a single exposure, one has to specify the response function g. The simplest way is to use the stochastic model which in radiobiological literature is referred to as the multihit-one target model (see Chapter 1). According to this model, one expects the distribution of the number of (irreparable) lesions induced in a cell exposed to a dose D to be Poisson with the parameter xD. Let m be the maximal number of such lesions (hits) a cell can bear without being killed. Then

$$g(x, D) = f_m(xD) \quad,$$

where

$$f_m(t) := e^{-t} \sum_{k=0}^{m} \frac{t^k}{k!}, \quad m \in \mathbb{Z}_+ \quad, \tag{2.3}$$

is the so-called multihit function. For an arbitrary irradiation scheme $\mathscr{D} = (D_1, \ldots, D_n)$, Formula 2.2 yields

$$g(x, \mathscr{D}) = e^{-xD} \prod_{i=1}^{n} \sum_{k=0}^{m} \frac{(xD_i)^k}{k!} \quad, \tag{2.4}$$

where $D = \sum_{i=1}^{n} D_i$.

In the sequel, we need one more assumption that complements the three preceding ones and concerns the parameters used in the hit and target model.

2.2.4   Assumption 4

It is the parameter x that describes variability of cells with respect to their radiosensitivity. The critical number m of hits is nonrandom, and it is the same for normal and malignant cells.

It should be noted that the multihit functions are merely approximations to the unknown true survival probabilities. Therefore, the following natural problem arises: to find the precise upper bound

$$\Phi_{\mathscr{F}} = \sup_{f \in \mathscr{F}} \int_0^{\infty} f(x)d(\mu_0 - \mu_1)(x) \tag{2.5}$$

for a class $\mathscr{F}$, depending on Δ and containing all admissible response functions satisfying minimal requirements which reflect only general properties of survival probabilities. If the value of $\Phi_{\mathscr{F}}$ is small, then this would suggest that the search for an optimal irradiation regimen is futile.

Problem 2.5 is solved in the following cases: (1) for all possible irradiation schemes ($\Delta = \Delta_0$); (2) for irradiation schemes with bounded total dose ($\Delta = \tilde{\Delta}(D)$).

In case 1, it is natural to consider $\mathcal{F} = W$, where W is the class of all nonincreasing absolutely continuous functions f on $\mathbb{R}_+$ such that $f(0) = 1$ and $\lim_{x \to +\infty} f(x) = 0$. In case 2, the class $\mathcal{F}$ may be identified with a more narrow set L_M of all nonincreasing functions f on $\mathbb{R}_+$ satisfying the Lipschitz condition

$$|f(x) - f(y)| \leq M|x - y|, \quad x, y \in \mathbb{R}_+ \quad , \tag{2.6}$$

and such that $f(0) = 1$, $\lim_{x \to +\infty} f(x) = 0$.

Here are the main arguments supporting our choice of these two classes of functions:

1. Every response function of type 2.4 belongs to the class L_M with
   $$M = De^{-m}\frac{m^m}{m!}.$$
2. It is possible to modify Function 2.4 and thus to take into account the contribution of intracellular repair processes to the resultant response of a cell to irradiation. The modified survival probability again appears to belong to the class L_M with the same M.
3. The set $\underset{M>0}{U} L_M$ is dense in W in the usual norm induced on W from the space of all absolutely continuous functions on $\mathbb{R}_+$ vanishing at infinity (see Appendix 1).
4. As some biological examples show, even the crude bound Φ_W is nearly attained in some cases.

The derivation of upper bounds (Equation 2.5) will be presented in Section 2.3. As a by-product, we obtain in this section a family of new metrics on the space of probabilistic measures on $\mathbb{R}$.

2.3 PRECISE UPPER BOUNDS FOR THE EFFICIENCY FUNCTIONAL OVER CLASSES OF ADMISSIBLE RESPONSE FUNCTIONS

In this section, we aim to calculate the values for two natural classes $\mathcal{F}$ of response functions

$$\Phi_{\mathscr{F}} = \sup_{f \in \mathscr{F}} \Phi(f) \quad ,$$

where

$$\Phi(f) = \int_0^\infty f(x)d(\mu_0 - \mu_1)(x)$$

and μ_0, μ_1 are two probabilistic measures on $\mathbb{R}_+$ with the distribution functions F_0, F_1.

A general assumption that cell response function $g(x, \mathscr{D})$ has the intensity with respect to radiosensitivity x and is determined uniquely by this intensity leads us to the class W of all absolutely continuous functions f on $\mathbb{R}_+$ such that $f(0) = 1$ and $\lim\limits_{x \to +\infty} f(x) = 0$. We remind that the derivative f' of a function $f \in W$ is defined almost everywhere (a.e.) with respect to the Lebesgue measure, $f' \leq 0$ a.e., and $\int_0^\infty f'(x)dx = -1$.

2.3.1   Theorem 1

$$\Phi_W = \sup_{x \in \mathbb{R}_+} (F_0 - F_1)(x) \quad .$$

Proof — Denote hereafter $\mu: = \mu_0 - \mu_1$ and $F: = F_0 - F_1$. Clearly, $F(0) = \lim\limits_{x \to +\infty} F(x) = 0$. Set $A = \sup\limits_{x \in \mathbb{R}_+} F(x)$. Integrating by parts, we obtain

$$\Phi(f) = \int_0^\infty f(x)d\mu(x) = -\int_0^\infty f'(x)F(x)dx, \quad f \in W \quad . \tag{3.1}$$

Hence, $\Phi_W \leq A$. While proving the converse inequality, we may assume that $A > 0$. Define the sets

$$E_n = \left\{ x \in \mathbb{R}_+ : F(x) \geq A - \frac{1}{n} \right\}, \quad n \in \mathbb{N} \quad .$$

Obviously, there is an integer N such that $0 < \text{mes}E_n < +\infty$ for all $n \geq N$, where mes stands for the Lebesgue measure. The functions

$$f_n(x) = 1 - (\text{mes}E_n)^{-1} \int_0^x \chi_{E_n}(t)dt, \qquad n \geq N \ , \tag{3.2}$$

belong to W as by 3.1 $\Phi(f_n) \geq A - \dfrac{1}{n}$ (here χ_E is the characteristic function of a set E). Thus, $\Phi_W \geq A$, and the proof is completed.

Remark — Introduce the set $E = \{x \in \mathbb{R}_+ : F(x) = \sup\limits_{t \in \mathbb{R}_+} F(t)\}$. In the case when mesE > 0, the supremum of Φ over W is attained at the function

$$f^*(x) = 1 - (\text{mes}E)^{-1} \int_0^x \chi_E(t)dt \ . \tag{3.3}$$

On the contrary, if mesE $= 0$, then the supremum Φ_W is not attained, but for the functions 3.2, we obviously have $\Phi(f_n) \to \Phi_W$ as $n \to \infty$. To prove the former, suppose that

$$A = \sup_{f \in W} \Phi(f) = \Phi(g)$$

for some $g \in W$. Hence,

$$-\int_0^\infty g'(x)F(x)dx = \sup_{t \in \mathbb{R}_+} F(t) = A \ .$$

Since $\int_0^\infty g'(x)dx = -1$, this implies

$$\int_0^\infty g'(x)(F(x) - A)dx = 0 \ .$$

The integrand is nonnegative. Hence, $g'(x)(A - F(x)) = 0$ a.e., that is, $g'(x) = 0$ a.e. on $\mathbb{R}_+ \setminus E$. From the condition mesE $= 0$, it follows that $g'(x) = 0$ a.e. Therefore, $g \equiv 1$, which is impossible because $\lim\limits_{x \to +\infty} g(x) = 0$.

Consider the most important particular case when there is the unique point $x_0 \in \mathbb{R}_+$ such that $\sup\limits_{x \in \mathbb{R}_+} F(x) = \max\{F(x_0-), F(x_0+)\}$. If we suppose additionally that the function F is continuous at the point x_0, then $E = \{x_0\}$ and $\Phi = \Phi(g_0)$, where the function

$$g_0(x) = \begin{cases} 1, & 0 \le x < x_0 \\ \dfrac{1}{2}, & x = x_0 \\ 0, & x > x_0 \end{cases} \tag{3.4}$$

does not belong, indeed, to W. Note also that the value $g_0(x_0)$ may be chosen arbitrarily.

It is important to underline here that the function g_0 can be approximated by the hit functions in the following sense: if $D_m = m/x_0$, then

$$f_m(xD_m) \rightarrow g_0(x) \text{ as } m \rightarrow \infty \text{ for all } x \in \mathbb{R}_+ \tag{3.5}$$

(for proof, see Appendix 2).

The function g_0 describes the deterministic type of the cellular response to irradiation. This means that cells with radiosensitivity less than x_0 survive with the probability 1 while those with radiosensitivity greater than x_0 die with the probability 1. This type of cell ordering with respect to their radiosensitivity is a characteristic feature of the so-called interphase death of cells exposed to high doses of ionizing radiation. Relation 3.5 shows that the value of the treatment efficiency for the "threshold" response function 3.4 and for sufficiently large values of m would be quite close to the upper bound Φ_W, if dose $D = m/x_0$ is applied. Thus, if one imposes no restriction on the dose of irradiation, the bound Φ_W is the best possible.

The intensity of the real cellular response function $g(x, \mathscr{D})$ with respect to radiosensitivity x is bounded. Its upper bound coincides with the (least) Lipschitz constant M of the function $g(x, \mathscr{D})$ with respect to x, if g is absolutely continuous. Actually, from the formula

$$g(x) = g(0) + \int_0^x g'(t)dt, \quad x \in \mathbb{R}_+ \quad ,$$

we see that, for all $x, y \in \mathbb{R}_+$,

$$|g(x) - g(y)| = \left| \int_x^y g'(t)dt \right| \le |x - y| \underset{t \in \mathbb{R}_+}{\text{essup}} |g'(t)| \quad .$$

Hence $M \le \underset{t \in \mathbb{R}_+}{\text{essup}} |g'(t)|$. Conversely, if there exists the derivative $g'(t)$

at a point $t \in \mathbb{R}_+$, then from the inequality

$$|g(t + h) - g(t)| \leq M|h|$$

we conclude that $|g'(t)| \leq M$. Thus,

$$\operatorname*{essup}_{t \in \mathbb{R}_+} |g'(t)| \leq M \quad .$$

In the case when irradiation regimens with limited total dose are considered, this observation makes it natural to introduce the class L_M of all nonincreasing functions f on $\mathbb{R}_+$ such that $f(0) = 1$, $\lim_{x \to +\infty} f(x) = 0$, and

$$|f(x) - f(y)| \leq M|x - y| \quad \text{for all } x, y \in \mathbb{R}_+ \quad .$$

The hit and target hypothesis enables explicit calculation of the Lipschitz constant M. In the case of single exposure to a dose D, we have

$$\sup_{x \in \mathbb{R}_+} \left| \frac{\partial}{\partial x} (f_m(xD)) \right| = a_m D \quad , \tag{3.6}$$

where f_m is the m-hit function 2.3 with the derivative $f'(t) = -e^{-t} \dfrac{t^m}{m!}$, and

$$a_m = \sup_{t \in \mathbb{R}_+} |f'_m(t)| = |f'_m(m)| = \begin{cases} 1, & m = 0 \\ e^{-m} \dfrac{m^m}{m!}, & m \geq 1 \end{cases} \quad .$$

Note that the sequence $\{a_m\}_{m \in \mathbb{Z}_+}$ is decreasing. Observe also that by the Stirling formula

$$a_m \leq (2\pi m)^{-1/2}$$

and, moreover, that

$$\lim_{m \to \infty} (2\pi m)^{1/2} a_m = 1 \quad .$$

It follows from 3.6 that in the multifractional case we have for the function 2.4 with $\mathcal{D} \in \Delta(D)$ the following estimate

$$\sup_{x \in R_+} \left| \frac{\partial}{\partial x} (g(x, \mathcal{D})) \right| \leq a_m D \quad . \tag{3.7}$$

Thus, response functions 2.4 satisfy the Lipschitz condition with the constant

$$M := \sup_{\mathcal{D} \in \Delta(D)} \sup_{x \in R_+} \left| \frac{\partial}{\partial x} (g(x, \mathcal{D})) \right| = a_m D \quad .$$

We will now show that this conclusion is also valid when we take into account contribution of the process of damage repair to the resulting response function. Consider a simple model in which the recovery of cells from radiation damage is provided by special repairing units (we call them *reparons*) operating at the genomic level. It is assumed that each intact reparon is capable of repairing one lesion caused by a hit into the cell target. Reparons are also subject to radiation damage. Suppose a cell with radiosensitivity x possesses r reparons and is exposed to a dose D. Denote by I the number of hits and by J the number of damaged reparons. Random variables I and J are considered to be independent; hence, the survival probability of the cell is equal to

$$g(x, D) = P(I + J \leq m + r) = \sum_{\substack{i+j \leq m+r \\ i \geq 0, \ 0 \leq j \leq r}} P(I = i)P(J = j) =$$

$$\sum_{j=0}^{r} P(J = j) \sum_{i=0}^{m+r-j} P(I = i) = \sum_{j=0}^{r} P(J = j)f_{m+r-j}(xD) \quad . \tag{3.8}$$

Further, since the sequence $\{a_m\}_{m \in Z_+}$ is decreasing, we have via 3.6

$$\sup_{x \in R_+} \left| \frac{\partial}{\partial x} (g(x, D)) \right| \leq D \sum_{j=0}^{r} P(J = j)a_{m+r-j} \leq$$

$$a_m D \sum_{j=0}^{r} P(J = j) = a_m D \quad .$$

This implies that 3.7 remains valid in the more general situation specified above.

Recall the following definition. Let $(X, \mathcal{A}, \mu)$ be a measure space with a nonnegative measure μ. The measure μ is called *nonatomic*, if for every set $A \in \mathcal{A}$ with $\mu A > 0$, the function $B \mapsto \mu B$ for $B \in \mathcal{A}$, $B \subset A$, takes all intermediate values between 0 and μA.

An explicit formula for the bound Φ_{LM} is based on the following result.[25]

Lemma 1 — Let $(X, \mathcal{A}, \mu)$ be a measurable space with a nonatomic measure μ such that $\mu X = +\infty$, and F be a measurable real function on X. For a, b > 0, define the set of measurable functions on X

$$G_{a,b} = \left\{ g : 0 \leq g \leq a \text{ a.e. and } \int_X g d\mu = b \right\} \quad ,$$

and denote $c = \sup \left\{ x : \mu(\{F \geq x\}) \geq \dfrac{b}{a} \right\}$.

1. If $\mu(\{F \geq c\}) \geq \dfrac{b}{a}$, then

$$\sup_{g \in G_{a,b}} \int_X g F d\mu = a \int_E F d\mu \tag{3.9}$$

for a set $E \in \mathcal{A}$ such that $\mu E = \dfrac{b}{a}$ and $\{F > c\} \subset E \subset \{F \geq c\}$.

2. If $\mu(\{F \geq c\}) < \dfrac{b}{a}$, then

$$\sup_{g \in G_{a,b}} \int_X g F d\mu = a \int_{\{F \geq c\}} F d\mu + c(b - a\mu(\{F \geq c\})) \quad . \tag{3.10}$$

Proof — First, let us consider the case $\mu(\{F \geq c\}) \geq \dfrac{b}{a}$. Note that $\mu(\{F > c\}) \leq \dfrac{b}{a}$. To see this, observe that by the definition of the number c we have, for every $c' > c$, $\mu(\{F \geq c'\}) < \dfrac{b}{a}$. Now choose a sequence $\{c_n\}_{n \in \mathbb{N}}$ such that $c_{n+1} < c_n$ for all $n \in \mathbb{N}$ and $c_n \to c$ as $n \to \infty$. Then, $\{F > c\} = \bigcup_n \{F \geq c_n\}$, where the sets in the right-hand side are nondecreasing, and the required inequality follows from the continuity of measures.

Since μ is a nonatomic measure, we can choose such $E \in \mathcal{A}$ that

$$\mu E = \frac{b}{a} \quad \text{and} \quad \{F > c\} \subset E \subset \{F \geq c\} \quad . \qquad (3.11)$$

We claim that

$$\int_X gFd\mu \leq a \int_E Fd\mu$$

for all $g \in G_{a,b}$. From 3.11, we have

$$\int_X gFd\mu = \int_E gFd\mu + \int_{X \setminus E} gFd\mu \leq \int_E gFd\mu + c \int_{X \setminus E} gd\mu =$$

$$\int_E gFd\mu + c\left(b - \int_E gd\mu\right) = \int_E gFd\mu + c \int_E (a - g)d\mu \leq$$

$$\int_E gFd\mu + \int_E (a - g)Fd\mu = a \int_E Fd\mu \quad .$$

The function $g = a\chi_E$ belongs to $G_{a,b}$ and $\int_X a\chi_E Fd\mu = a \int_E Fd\mu$. Thus, the statement of the lemma, for the case under study, is established.

Consider now the case $\mu(\{F \geq c\}) < \frac{b}{a}$. Denoting $E = \{F \geq c\}$ and repeating the first three transitions in the above estimates, we have for all $g \in G_{a,b}$

$$\int_X gFd\mu \leq \int_E gFd\mu + c\left(b - \int_E gd\mu\right) = \int_E g(F - c)d\mu + cb \leq$$

$$a \int_E (F - c)d\mu + cb = a \int_E Fd\mu + c(b - a\mu E) \quad .$$

Thus,

$$\sup_{g \in G_{a,b}} \int_X gFd\mu \leq a \int_{\{F \geq c\}} Fd\mu + c(b - a\mu(\{F \geq c\})) \quad .$$

To obtain the inverse inequality, take such a sequence $\{c_n\}_{n \in \mathbb{N}}$ that $c_n < c_{n+1}$ for all n and $c_n \to c$ as $n \to \infty$. We will show that $\mu(\{F \geq c_n\}) = +\infty$ for all n. In fact, suppose this statement fails for some $n = n_0$. The definition of c implies that $\mu(\{F \geq c_n\}) \geq \dfrac{b}{a}$ for all

n. Since $\{F \geq c\} = \bigcap\limits_{n \geq n_0} \{F \geq c_n\}$, $\{F \geq c_{n+1}\} \subset \{F \geq c_n\}$ for all n, and

$\mu(\{F \geq c_{n_0}\}) < +\infty$, we have, via continuity of finite measures at zero,

the inequality $\mu(\{F \geq c\}) \geq \dfrac{b}{a}$, which contradicts the assumption stated

above.

The measure μ is nonatomic; hence, we can choose for each n a

set $E_n \in \mathcal{A}$ with the properties $\mu E_n = \dfrac{b}{a}$, $E \subset E_n \subset \{F \geq c_n\}$. For the

function $g_n = a\chi_{E_n} \in G_{a,b}$, we have

$$\int_X g_n F d\mu = a \int_{E_n} F d\mu = a \int_E F d\mu + a \int_{E_n \setminus E} F d\mu \geq$$

$$a \int_E F d\mu + ac_n \left(\frac{b}{a} - \mu E\right) = a \int_E F d\mu +$$

$$c(b - a\mu E) - (c - c_n)(b - a\mu E) \quad .$$

Hence,

$$\sup_{g \in G_{a,b}} \int_X g F d\mu \geq a \int_E F d\mu + c(b - a\mu E) \quad ,$$

and this completes the proof of the lemma.

Remark 1 — Though the set E is not necessarily unique, the value

$\int_E F d\mu$ does not depend on the particular choice of E.

Remark 2 — In the case $\mu(\{F \geq c\}) < \dfrac{b}{a}$, we have $\mu(\{F \geq c'\}) =$

$+\infty$ for all $c' < c$.

Now we are in a position to formulate the main result of Section
2.3.

2.3.2 Theorem 2

Denote $c = \sup \left\{ x : \text{mes} \left(\{F_0 - F_1 \geq x\}\right) \geq \dfrac{1}{M} \right\}$.

1. If $\text{mes}(\{F_0 \geq F_1\}) \geq \dfrac{1}{M}$, then

$$\Phi_{L_M} = M \int_E (F_0 - F_1)(x)dx \quad , \tag{3.12}$$

where $\text{mes}E = \dfrac{1}{M}$ and $\{F_0 - F_1 > c\} \subset E \subset \{F_0 - F_1 \geq c\}$.

2. If $\text{mes}(\{F_0 \geq F_1\}) < \dfrac{1}{M}$, then

$$\Phi_{L_M} = M \int_{\{F_0 \geq F_1\}} (F_0 - F_1)(x)dx \quad . \tag{3.13}$$

Proof — Observe that the set of all functions $-f'$ for $f \in L_M$ coincides with the set $G_{M,1}$ [see Lemma 1 where $(X, \mathscr{A}, \mu)$ is $\mathbb{R}_+$ with the Lebesgue measure]. Suppose $\text{mes}(\{F_0 \geq F_1\}) < \dfrac{1}{M}$. Then $c = 0$, since $\lim\limits_{x \to +\infty} (F_0 - F_1) (x) = 0$. Using 3.1 and 3.10, we obtain 3.13. Formula 3.12 is derived straight from 3.9 by means of the same argument.

Remark 1 — In the case $\text{mes}(\{F_0 \geq F_1\}) \geq \dfrac{1}{M}$, the supremum of the functional Φ over L_M is attained at function 3.3 with the set E indicated in Theorem 2. In contrast, if $\text{mes}(\{F_0 \geq F_1\}) < \dfrac{1}{M}$, then the upper bound Φ_L is not attained. But a sequence of functions $\{f_n\}_{n \in \mathbb{N}}$ in L_M, such that $\Phi(f_n) \to \Phi_{L_M}$ as $n \to \infty$, can be pointed out explicitly.

To prove the first of these assertions, assume that Φ_{L_M} is attained. Then in view of 3.1 and 3.13, there exists a function g on $\mathbb{R}_+$ such that $0 \leq g \leq M$ a.e., $\displaystyle\int_0^\infty g(x)dx = 1$, and

$$M \int_0^\infty F(x)\chi_{\{F \geq 0\}}(x)dx = \int_0^\infty F(x)g(x)dx \quad ,$$

where $F = F_0 - F_1$. Hence, we have

$$\int_0^\infty F(x)(M\chi_{\{F \geq 0\}}(x) - g(x))dx = 0 \quad .$$

Observe that the function $F(x)$ $(M\chi_{\{F \geq 0\}}(x) - g(x))$ is nonnegative a.e. Consequently, it is equal to 0 for almost all $x \in R_+$. So, $g(x) = 0$ for almost all $x \in R_+$ such that $F(x) < 0$, and we finally obtain the estimate

$$\int_0^\infty g(x)dx = \int_{\{F \geq 0\}} g(x)dx \leq M \, \text{mes}(\{F \geq 0\}) < 1$$

which is a contradiction.

Now define the following sequence of functions

$$f_n(x) = 1 - (\text{mes}U_n)^{-1} \int_0^x \chi_{U_n}(t)dt, \quad n \in \mathbb{N} \quad,$$

where $\{U_n\}_{n \in \mathbb{N}}$ is a sequence of subsets in R_+ defined by the conditions $U_n = \{F_0 \geq F_1\} \cup V_n$, $V_n \subset \{F_0 < F_1\}$, $\text{mes}V_n = \dfrac{1}{M} - \text{mes}(\{F_0 \geq F_1\})$, and $\inf V_n \to +\infty$ as $n + \infty$. Note that

$$\Phi(f_n) = -\int_0^\infty F(x)f_n'(x)dx = (\text{mes}U_n)^{-1} \int_{U_n} F(x)dx \quad,$$

and recall that in the case $\text{mes}(\{F \geq 0\}) < \dfrac{1}{M}$

$$\Phi_{L_M} = M \int_{\{F \geq 0\}} F(x)dx \quad.$$

Since $\text{mes}U_n = \dfrac{1}{M}$, we have (see definition of U_n)

$$|\Phi_{L_M} - \Phi(f_n)| = M \left| \int_{V_n} F(x)dx \right| \leq M \sup_{x \in V_n} |F(x)| \text{mes}V_n \leq \sup_{x \in V_n} |F(x)| \quad.$$

Relations $\inf V_n \to +\infty$ as $n \to \infty$ and $F(x) \to 0$ as $x \to +\infty$ imply

$$\Phi(f_n) \to \Phi_{L_M}, \quad n \to \infty \quad.$$

Remark 2 — If $M_1 \leq M_2$, then $L_{M_1} \subset L_{M_2}$. Hence, the function $M \mapsto \Phi_{L_M}$ is nondecreasing. We remind that in the hit and target model $M = a_m D$, where the sequence $\{a_m\}_{m \in Z_+}$ is decreasing. Thus, the Lipschitz bound Φ_{L_M} increases with respect to D and decreases with respect to m. Therefore, it takes its greatest value for D being the maximal tolerant dose and for m = 0.

Remark 3 — Suppose the distributions μ_0, μ_1 have finite expectations $\bar{\mu}_0$, $\bar{\mu}_1$, and μ_0 stochastically dominates μ_1 (i.e., $F_0(x) \geq F_1(x)$ for all $x \in \mathbb{R}_+$). Then for $0 < M \leq 1$, we obtain

$$\Phi_{L_M} \leq \Phi_{L_1} = \int_E (F_0 - F_1)(x)dx \leq \int_0^\infty (F_0 - F_1)(x)dx =$$

$$\int_0^\infty (\bar{F}_1 - \bar{F}_2)(x)dx = \bar{\mu}_1 - \bar{\mu}_0 \quad ,$$

where $\bar{F}_i = 1 - F_i$, i = 0, 1.

The inclusion $L_m \subset W$ implies $\Phi_{L_M} \geq \Phi_W$. Moreover, it follows from Theorem 2 that $\Phi_{L_M} \rightarrow \Phi_W$ as $M \rightarrow +\infty$. It is easy to show that $\Phi_{L_M} = \Phi_W$, if and only if $\Phi_W = 0$ or

$$\text{mes}\left(\left\{x : (F_0 - F_1)(x) = \sup_{t \in \mathbb{R}_+} (F_0 - F_1)(t)\right\}\right) \geq \frac{1}{M} \quad .$$

To prove this statement, suppose that $\Phi_{L_M} = \Phi_W$. Denote $A = \Phi_W = \sup_{t \in \mathbb{R}_+} F(t)$. Consider case 1, i.e. $\text{mes}(\{F \geq 0\}) \geq \frac{1}{M}$. Then $\Phi_{L_M} = \Phi_W$ means that

$$M \int_E F dx = A$$

or, via the equality $\text{mes}E = \frac{1}{M}$, that

$$M \int_E (A - F)dx = 0 \quad .$$

Hence, F = A on E a.e., and

$$\text{mes}(\{F = A\}) \geq \text{mes}E = \frac{1}{M} \quad .$$

In case 2, $\text{mes}G < \dfrac{1}{M}$, where $G = \{F \geq 0\}$, we have $A = \Phi_W = \Phi_{LM}$

$= \displaystyle\int_G Fdx \leq M \, A \, \text{mes}G$. Since $A \geq 0$, this implies necessarily that $A = 0$. Now let us prove the reverse implication. Suppose $\Phi_W = 0$. It follows from the relation $\lim\limits_{x \to +\infty} F(x) = 0$ and from the definition of number c that $c \geq 0$. Hence, $\Phi_{LM} \geq 0$ (see 3.12 and 3.13). So, the inequality $\Phi_{LM} \leq \Phi_W$ implies $\Phi_{LM} = \Phi_W = 0$. Now suppose that $\text{mes}(\{F = A\}) \geq \dfrac{1}{M}$. Then, $c = A$ and $E \subset \{F = A\}$. Thus,

$$\Phi_{LM} = M \int_E Fdx = MA\text{mes}E = A \quad .$$

The bound Φ_{LM} is worth comparing also with the bound $\Phi_{Lip(M)}$, where Lip(M) is the set of *all* functions on $\mathbb{R}_+$ satisfying the Lipschitz condition 3.6. The value

$$\Phi_{Lip(M)} = \sup_{f \in Lip(M)} \int_0^\infty f(x)d(\mu_0 - \mu_1)(x)$$

is known as the Kantorovich-Rubinstein distance between measures μ_0 and μ_1.[34] It follows from 3.1 that

$$\Phi_{Lip(M)} = M \int_0^\infty |(F_0 - F_1)(x)|dx \quad ,$$

and the extremal function $f^* \in Lip(M)$ is defined by means of the equality $f^{*'}(x) = M \, \text{sign} \, (F_1 - F_0)(x)$. Indeed, $\Phi_{LM} \leq \Phi_{Lip(M)}$ while the equality here takes place, if and only if $F_0(x) \geq F_1(x)$ for all $x \in \mathbb{R}_+$ and $\text{mes}(\{F_0 > F_1\}) \leq \dfrac{1}{M}$.

To prove this, suppose that $F_0 \geq F_1$ and $\text{mes}(\{F_0 > F_1\}) \leq \dfrac{1}{M}$. We claim that

$$\int_E Fdx = \int_{\{F>0\}} Fdx \quad .$$

To make sure that this equality is valid, consider the following two cases. (1) $c = 0$. In this case, the above equality follows from the inclusions $\{F > 0\} \subset E \subset \{F \geq 0\}$ (see definition 3.11 of set E). (2) $c > 0$. In this case, we have $E \subset \{F \geq c\} \subset \{F > 0\}$, but since $\text{mes}E = \dfrac{1}{M}$ and $\text{mes}(\{F > 0\}) \leq \dfrac{1}{M}$ by the hypothesis, we obtain $\text{mes}(\{F > 0\} \backslash E) = 0$. Consequently, $\displaystyle\int_E F dx = \int_{\{F>0\}} F dx$. Now we have

$$\Phi_{\text{Lip}(M)} = M \int_0^\infty |F| dx = M \int_{\{F>0\}} F dx = M \int_E F dx = \Phi_{\text{LM}} \ .$$

Conversely, let us assume that $\Phi_{\text{LM}} = \Phi_{\text{Lip}(M)}$. Since $c \geq 0$, we see that $E \subset \{F \geq 0\}$. Hence, in any case $\Phi_{\text{LM}} \leq M \displaystyle\int_{\{F \geq 0\}} F dx$, but the latter does not exceed $M \displaystyle\int_0^\infty |F| dx = \Phi_{\text{Lip}(M)}$. Therefore, if $\Phi_{\text{LM}} = \Phi_{\text{Lip}(M)}$, then $\displaystyle\int_{\{F \geq 0\}} F dx = \int_0^\infty |F| dx$. This implies $\text{mes}(\{F<0\}) = 0$, i.e., via the left continuity of F, $F(x) \geq 0$ for all $x \in R_+$. Now we will prove that $\text{mes}(\{F > 0\}) \leq \dfrac{1}{M}$. If $c = 0$, this follows from the definition of the set E. Let $c > 0$. As it was shown earlier, $E \subset \{F > 0\}$. The equality $\Phi_{\text{LM}} = \Phi_{\text{Lip}(M)}$ yields

$$\int_E F dx = \int_{\{F>0\}} F dx \ .$$

This implies $\text{mes}(\{F > 0\} \backslash E) = 0$, i.e., $\text{mes}(\{F > 0\}) = \text{mes}E = \dfrac{1}{M}$.

The Lipschitz bound Φ_{LM} gives rise to a family of new metrics on the space $\mathscr{P}$ of probability measures on $\mathbb{R}$.

Let F be a real measurable function on $\mathbb{R}$ such that $\lim\limits_{x \to \pm\infty} F(x) = 0$. For $r > 0$, define

$$c_r := \sup\{x : \text{mes}(\{F \geq x\}) \geq r\}$$

and

$$\psi_r(F) := \dfrac{1}{r} \int_{E_r} F(x) dx \ ,$$

where $mesE_r = r$, $\{F > c_r\} \subset E_r \subset \{F \geq c_r\}$, if $mes(\{F \geq 0\}) \geq r$, and $E_r = \{F \geq 0\}$, if $mes(\{F \geq 0\}) < r$. Set finally

$$\xi_r(\mu_0, \mu_1) := \max\{\psi_r(F_0 - F_1), \psi_r(F_1 - F_0)\}, \quad \mu_0, \mu_1 \in \mathscr{P} \quad ,$$

where F_0 and F_1 are the corresponding distribution functions.

Applying Lemma 1, we see that every ξ_r is a metric on $\mathscr{P}$. Moreover, according to Remark 2 related to Theorem 2, the family of metrics $\{\xi_r\}_{r>0}$ is nonincreasing with respect to r. These metrics "soften" the uniform metric

$$\xi_0(\mu_0, \mu_1) := \sup_{x \in \mathbb{R}} |(F_0 - F_1)(x)|$$

which appears to be their limit case:

$$\lim_{r \to 0+} \xi_r(\mu_0, \mu_1) = \xi_0(\mu_0, \mu_1), \quad \mu_0, \mu_1 \in \mathscr{P} \quad .$$

To show this, we observe that the inequality $\xi_r \leq \xi_0$ is an immediate consequence of the definition of ξ_r. Hence, $\lim_{r \to 0+} \xi_r \leq \xi_0$. For proof of the converse inequality, we may assume that $\xi_0 = A > 0$, where $A = \sup_{t \in \mathbb{R}_+} F(t)$, and $F = F_0 - F_1$, as before. Then $mes(\{F \geq 0\}) > 0$. For $r \leq mes(\{F \geq 0\})$, we have $mesE_r = r$ and $E_r \supset \{F > c_r\}$. Therefore,

$$\xi_r = \frac{1}{r} \int_{E_r} F dx \geq c_r \quad .$$

It follows from the definition of c_r that it is nondecreasing with respect to r. Denote $c = \lim_{r \to 0+} c_r$. For a given $\epsilon > 0$, there exists a point $x \in \mathbb{R}$ such that $F(x) > A - \epsilon$. The function F is left continuous; hence, there is $\delta > 0$ such that $F(y) \geq A - \delta$ for $y \in (x - \delta, x)$. Therefore, $c \geq c_\delta \geq A - \epsilon$. This implies via arbitrariness of ϵ that $c \geq A$, and we have finally

$$\lim_{r \to 0+} \xi_r \geq c \geq \xi_0 \quad .$$

Table 6
Variants of Pairs of Gamma Distributions Representing Radiosensitivity of Normal and Neoplastic Cells

Variant No.	$\bar{x}_0$	v_0	$\bar{x}_1$	v_1
1	0.2	0.25	0.4	0.25
2	0.2	0.5	0.4	0.25
3	0.4	0.5	0.4	0.25
4	0.2	0.25	0.4	0.5
5	0.4	0.25	0.4	0.5
6	0.1	0.25	0.5	0.25

From Hanin, L. G., Rachev, S. T., and Yakovlev, A. Yu., *Adv. Appl. Probab.*, 25, 1, 1993. With permission.

2.4 NUMERICAL EXAMPLE

Both optimal (or almost optimal) response functions discussed in Section 2.3 do not seem to be reproducible in reality. This leads us to the question: What gain in therapeutical efficiency do they give in comparison with real response functions? We get some idea of this gain by means of numerical example.

For measures μ_0 and μ_1, we use gamma distributions with densities p_{α_i,β_i}, $i = 0, 1$, where

$$p_{\alpha,\beta}(x) = \beta^\alpha(\Gamma(\alpha))^{-1}x^\alpha e^{-\beta x}\chi_{\mathbb{R}_+}(x), \quad \alpha, \beta > 0 \ .$$

Every gamma distribution is determined uniquely by its mean value $\bar{x} = \alpha/\beta$ and its variation coefficient $v = 1/\sqrt{\alpha}$. The variants of pairs $\bar{x}_0, v_0$ and $\bar{x}_1, v_1$ chosen throughout for calculations are given in Table 6. The graphs of the corresponding functions $F = F_0 - F_1$ are shown in Figure 9.

For every pair of gamma distributions μ_0, μ_1 and for $m = 1, 2, 3, 19$, the optimal single dose D_{opt} specified by the condition

$$\int_0^\infty f_m(xD)d(\mu_0 - \mu_1)(x) \to \max$$

and the corresponding real optimal efficiency

$$\Phi(D_{opt}) = \int_0^\infty f_m(xD_{opt})d(\mu_0 - \mu_1)(x)$$

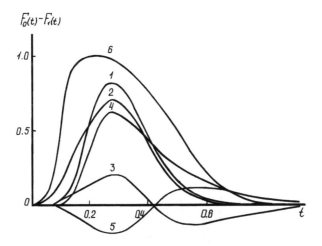

Figure 9
Plots of the function $F_0 - F_1$ corresponding to the combinations of parameters presented in Table 6.

were calculated. Then the theoretical bounds Φ_{L_M} with $M = a_m D_{opt}$ and Φ_W were computed. The results are presented in Table 7. They reveal comparatively small difference between the maximal multihit efficiency and its Lipschitz theoretical bound.

2.5   AN ESTIMATE FOR THE EFFICIENCY FUNCTIONAL IN THE CASE OF ARBITRARY DISTRIBUTIONS OF THE CRITICAL NUMBER OF HITS

In the preceding exposition, it was supposed that the value m of the hit parameter is the same for all cells, both normal and neoplastic.

In this section, we will omit this assumption and obtain an estimate of the efficiency functional for a pair of arbitrary distributions of the hit parameter m in normal and neoplastic tissues. For the sake of simplicity, we confine ourselves to the case of a single exposure.

Let $\rho_0 = \sum_{m=0}^{\infty} \alpha_m^0 \delta_m$ and $\rho_1 = \sum_{m=0}^{\infty} \alpha_m^1 \delta_m$ be the distributions of the critical value of unrepaired lesions, m, for normal and neoplastic cells, respectively. We use designations R_0 and R_1 for the corresponding distribution functions. Denote $\Delta(x, D)$ the difference between survival

Table 7

Theoretical Upper Bounds of the Efficiency Functional Φ Over Lipschitz Class L_M and Over Class W of Absolutely Continuous Functions Compared with the Maximal Single Dose Efficiency $\Phi(D_{opt})$

Variant No.	m = 1		m = 2		m = 3		m = 19		Φ_W
	$\Phi(D_{opt})$	Φ_{LM}	$\Phi(D_{opt})$	Φ_{LM}	$\Phi(D_{opt})$	Φ_{LM}	$\Phi(D_{opt})$	Φ_{LM}	
1	0.341	0.499	0.406	0.511	0.455	0.564	0.694	0.752	0.831
2	0.359	0.517	0.417	0.542	0.457	0.551	0.625	0.658	0.706
3	0.069	0.136	0.086	0.138	0.098	0.161	0.163	0.182	0.204
4	0.284	0.389	0.334	0.436	0.370	0.482	0.537	0.574	0.626
5	0.006	0.025	0.014	0.032	0.021	0.033	0.077	0.093	0.125
6	0.689	0.903	0.779	0.924	0.835	0.956	0.981	0.992	0.998

From Hanin, L. G., Rachev, S. T., and Yakovlev, A. Yu., *Adv. Appl. Probab.*, 25, 1, 1993. With permission.

probabilities of normal and neoplastic cells with the radiosensitivity x exposed to dose D:

$$\Delta(x, D) = \sum_{m=0}^{\infty} f_m(xD)\alpha_m^0 - \sum_{m=0}^{\infty} f_m(xD)\alpha_m^1 = \varphi(xD) \quad , \qquad (5.1)$$

where

$$\varphi(t) = \sum_{m=0}^{\infty} f_m(t)(\alpha_m^0 - \alpha_m^1) \quad .$$

Then we have for $t \geq 0$

$$\varphi(t) = \sum_{m=0}^{\infty} \left(e^{-t} \sum_{k=0}^{m} \frac{t^k}{k!} \right)(\alpha_m^0 - \alpha_m^1) = e^{-t} \sum_{k=0}^{\infty} \sum_{m=k}^{\infty} \frac{t^k}{k!} (\alpha_m^0 - \alpha_m^1) =$$

$$e^{-t} \sum_{k=0}^{\infty} \frac{t^{k+1}}{(k + 1)!} \sum_{j=0}^{k} (\alpha_j^1 - \alpha_j^0) = e^{-t} \sum_{k=0}^{\infty} \frac{t^{k+1}}{(k + 1)!} (R_1(k) - R_0(k)) \leq$$

$$e^{-t} \sum_{\{k:R_1(k)>R_0(k)\}} \frac{t^{k+1}}{(k + 1)!} (R_1(k) - R_0(k)) \leq$$

$$e^{-t} \left(\sup_{n \in \mathbb{N}} \frac{t^n}{n!} \right) \int_{\{R_1>R_0\}} (R_1 - R_0)(t)dt \quad .$$

It is easy to see that

$$\sup_{n \in \mathbb{N}} \frac{t^n}{n!} = \frac{t^{[t]}}{[t]!}, \quad t \geq 0 \quad ,$$

where [t] is 1 for $t \in (0, 1)$ and the largest integer $\leq t$ for $t \geq 1$. Introduce the function

$$h(t) = e^{-t} \frac{t^{[t]}}{[t]!}, \quad t \geq 0 \quad .$$

Then

$$\Delta(x, D) \leq h(xD) \int_{\{R_1>R_0\}} (R_1 - R_0)(t)dt \quad . \qquad (5.2)$$

Notice that 5.2 turns into equality, if and only if $\rho_0 = \delta_n$, $\rho_1 = \delta_{n-1}$, and [xD] = n for some $n \in \mathbb{N}$.

We point out also that inequality 5.2 holds, if the hit functions f_m in 5.1 are replaced by more general response functions $g_m(xD)$ of survival probability type such that

$$0 \leq g_{m+1}(t) - g_m(t) \leq h(t), \quad m \in \mathbb{Z}_+, \ t \geq 0 \ . \tag{5.3}$$

The purpose of this section is to evaluate the bound

$$\Phi_{\mathscr{F}} = \sup_{\{g_m\} \in \mathscr{F}} \Phi(\{g_m\}) \ ,$$

where

$$\Phi(\{g_m\}) = \int_{\mathbb{R}_+ \times \mathbb{Z}_+} g_m(xD) d(\mu_0 \times \rho_0 - \mu_1 \times \rho_1)(x, m)$$

and $\mathscr{F}$ is the set of all sequences $\{g_m\}_{m \in \mathbb{Z}_+}$ of functions $g_m \in L_{a_m}$ satisfying 5.3.

The functional Φ splits into two parts:

$$\Phi = \Phi_1 + \Phi_2 \ ,$$

where

$$\Phi_1(\{g_m\}) = \sum_{m=0}^{\infty} (\alpha_m^0 - \alpha_m^1) \int_0^{\infty} g_m(xD) d\mu_0(x) \ ,$$

$$\Phi_2(\{g_m\}) = \sum_{m=0}^{\infty} \alpha_m^1 \int_0^{\infty} g_m(xD) d(\mu_0 - \mu_1)(x) \ .$$

Combining 5.2 with the equality $\sup_{t \in \mathbb{R}} h(t) = e^{-1}$, we obtain the estimate

$$\Phi_1(\{g_m\}) \leq \int_0^{\infty} h(xD) d\mu_0(x) \int_{\{R_1 > R_0\}} (R_1 - R_0)(t) dt \leq$$

$$e^{-1} \int_{\{R_1 > R_0\}} (R_1 - R_0)(t) dt \ .$$

Taking into account the results of Section 2.3, we find that

$$\Phi_2(\{g_m\}) \leq \sup_m \int_0^\infty g_m(xD)d(\mu_0 - \mu_1)(x) \leq \sup_m \Phi_{L_{a_m}D} =$$

$$\Phi_{L_D} = \psi_{D^{-1}}(F_0 - F_1)$$

(for definition of Ψ_r, see also Section 2.3).

We have finally

$$\Phi_{\mathcal{F}} \leq e^{-1} \int_{\{R_1 > R_0\}} (R_1 - R_0)(t)dt + \Psi_{D^{-1}}(F_0 - F_1) \quad . \qquad (5.4)$$

Estimate 5.4 coincides with the Lipschitz bound for the populations of normal and malignant cells which are identically distributed with respect to m, and with bound 5.2 for the populations with the sensitivity $x = D^{-1}$. Thus, it is sharp in the sense indicated above.

3

Optimal Fractionation Problem

3.1 INTRODUCTION

In this chapter, we shall concentrate almost all our efforts on the problem of optimal fractionation of a given total dose, i.e., that consisting in finding the value

$$\Phi^*(\Delta(D)) := \sup_{\mathcal{D} \in \Delta(D)} \Phi(\mathcal{D}) \quad , \tag{1.1}$$

where $\Delta(D) := \{\mathcal{D} \in \Delta_0 : D_1 + \ldots + D_n = D\}$, $D > 0$, and the corresponding optimal fractionation schemes (see the more general problem 2.1 formulated in Section 2.2).

Problem 1.1 will be considered in the next section within the framework of the "hit and target" model. The main attention will be paid to the following natural questions.

1. Is the supremum in 1.1 attained? In other words, is it true that

$$\Phi^*(\Delta(D)) = \Phi^*(\Delta_N(D))$$

for some $N \in \mathbb{N}$ (the number N depending on the value D, of course)?

2. If the answer to the preceding question is yes, then does the optimal scheme $\mathcal{D}^* = (D_1^*, D_2^*, \ldots, D_{n^*}^*)$ appear to be uniform (i.e., $D_1^* = D_2^* = \ldots = D_{n^*}^*$) and is it possible to estimate the number n^* in this case?

If $m = 0$, then $g(x, \mathcal{D}) = e^{-xD}$ for all $\mathcal{D} \in \Delta(D)$, and the fractionation problem is trivial.

In another particular case m = 1, the complete solution can also be found. Namely, for every total dose D > 0, there always exists a uniform optimal fractionation with the number of treatments bounded from above and an estimate of this bound can be obtained.

On the contrary, for m ≥ 2 uniform fractionation schemes are not necessarily optimal. Still, in some cases they are, and it is worth finding an upper bound estimating *a priori* the number of fractions in optimal (or at least sufficiently good) uniform irradiation scheme. We give such an estimate in Section 3.2.

In the case m ≥ 2, the complete solution of the optimal fractionation problem remains unknown. To obtain an approximate solution, we reduce the problem to the case N = 2, i.e., to optimal partition of a given total dose into two fractions. This dichotomy problem turns out to be essentially algebraic and can be solved for all m ≥ 2 (see Section 3.2). By iterating the dichotomy process (i.e., applying it to parts of the total dose obtained in the previous steps), we get a diadic fractionation procedure with the efficiency hopefully approximating the optimal one. The numerical study presented in Section 3.2 supports this assumption.

As to the optimal fractionation problem without any restrictions on dose values ($\Delta = \Delta_0$), we managed to treat it only by means of computer-based experiments in the case of bounded number of fractions.

In Section 3.3, we introduce an alternative stepwise optimization procedure. It leads us to a noncommutative response function, i.e., the one depending on the order of elements in sequence $\mathscr{D} = (D_1, \ldots, D_n)$. This procedure is also investigated numerically.

It should be kept in mind that radiosensitivity distributions are unobservable. What we can know is their empirical estimates, those based on expert knowledge among them. This gives rise to the problem of stability of functional Φ, its extremal values and bounds with respect to perturbation of distributions μ_0 and μ_1. The corresponding estimates are given in Section 3.4.

3.2   OPTIMAL SCHEMES OF DOSE FRACTIONATION FOR THE MULTIHIT-ONE TARGET MODEL

In the sequel, response functions g(x, $\mathscr{D}$) in 1.1 from Chapter 2 will always be multihit functions.

In this section, we need a natural condition imposed on a pair of distributions μ_0, μ_1. It claims that single exposure of some doses provides positive therapeutic efficiency:

$$\Phi^*_{1,m} := \sup_{D>0} \int_0^\infty f_m(xD)d(\mu_0 - \mu_1)(x) > 0 \quad . \tag{2.1}$$

Remark 1 — If $\mu_0 \neq \mu_1$ and $F_1(x) \geq F_0(x)$ for all $x \in R_+$, then via 3.1 from Chapter 2, $\Phi(D) > 0$ for all $D > 0$; thus 2.1 holds. Conversely, if $F_0(x) \geq F_1(x)$ for all $x \in R_+$, then 2.1 is not true. Characterization of pairs μ_0, μ_1 satisfying 2.1 is a kind of stochastic dominance problem. A more general problem of considerable mathematical interest consists in describing all measures σ on R_+ such that their Laplace transform

$$z \to \int_0^\infty e^{-zx}d\sigma(x)$$

is nonnegative for every $z \in R_+$.

Remark 2 — Denote by $\Delta^*_{1,m}$ the set of all such $D \in R_+$ that $\Phi(D) = \Phi^*_{1,m}$. Obviously, $\Delta^*_{1,m}$ is nonempty.

Since $\Phi(0) = \lim\limits_{D \to +\infty} \Phi(D) = 0$ and in view of 2.1, it is also bounded away from zero and from infinity. Moreover, set $\Delta^*_{1,m}$ is finite. To show this, suppose the converse, then $\Delta^*_{1,m}$ has a limit point $D_0 > 0$. For every $D \in \Delta^*_{1,m}$, we have

$$\Phi'(D) = -\int_0^\infty e^{-xD} \frac{(xD)^m}{m!} xD\mu(x) = 0 \tag{2.2}$$

(we remind that $\mu = \mu_0 - \mu_1$). Thus, the set of zeros of the analytic function

$$F(z) = \int_0^\infty e^{-zx}x^{m+1}d\mu(x), \quad Re z > 0 \quad ,$$

has a limit point. Therefore, $F \equiv 0$, and by the uniqueness theorem for the Laplace transform, we obtain $x^{m+1} d\mu = 0$, i.e., $\mu = c\delta_0$. Since $\mu(R_+) = 0$, this implies $\mu = 0$, which leads to contradiction with 2.1.

Remark 3 — Value $\Phi^*_{1,m}$ is nondecreasing with respect to m. To prove this, take $D \in \Delta^*_{1,m}$. By 2.2, we obtain

$$\Phi^*_{1,m+1} \geq \int_0^\infty f_{m+1}(xD)d\mu(x) = \int_0^\infty f_m(xD)d\mu(x) = \Phi^*_{1,m} \quad .$$

So, if 2.1 holds for some m, then it is valid automatically for all greater values of the hit parameter.

Let D be a dose such that $\Phi(D) > 0$. This dose will be fixed as the total dose throughout the section.

Denote

$$\Phi^*(m; D) = \sup_{\mathscr{D} \in \Delta(D)} \int_0^\infty \prod_i f_m(xD_i)d(\mu_0 - \mu_1)(x)$$

and

$$\Phi^*_N(m; D) = \sup_{\mathscr{D} \in \Delta_N(D)} \int_0^\infty \prod_i f_m(xD_i)d(\mu_0 - \mu_1)(x) \quad . \qquad (2.3)$$

It is plain that $\Phi^*_N(m; D)$ is nondecreasing with respect to N and that

$$\Phi^*(m; D) = \lim_{N \to \infty} \Phi^*_N (m; D) \quad .$$

The supremum in 2.3 is attained; let $\Delta^*_N(m; D)$ be the set of the corresponding optimal schemes.

Let $m = 0$; in this case, $f_0(t) = e^{-t}$. Hence,

$$\Phi(\mathscr{D}) = \Phi(D) = \int_0^\infty e^{-xD}d(\mu_0 - \mu_1)(x)$$

is constant for all partitions $\mathscr{D}$ of the total dose D. This property holds only for $m = 0$ and thus can be utilized as the practical criterion for deciding whether or not $m = 0$.

Let $m = 1$; then $f_1(t) = e^{-t}(1 + t)$. We start with the dichotomy problem. Suppose $\mathscr{D} = (D_1, D_2)$, where $D_1 + D_2 = D$, and set $\lambda_1 = D_1/D$, $\lambda_2 = D_2/D$, $\lambda = \lambda_1\lambda_2$. Obviously, $\lambda_1, \lambda_2 \geq 0$, $\lambda_1 + \lambda_2 = 1$, and $0 \leq \lambda \leq 1/4$. We have

$$\Phi(\mathscr{D}) = \int_0^\infty e^{-xD}(1 + xD_1)(1 + xD_2)d\mu(x) = \int_0^\infty e^{-xD}(1 + xD)d\mu(x) +$$

$$\lambda \int_0^\infty e^{-xD}(xD)^2 d\mu(x) = \Phi(D) + \lambda I(D) \quad ,$$

where

$$I(D) = \int_0^\infty e^{-xD}(xD)^2 d\mu(x) \quad .$$

Hence, the irradiation efficiency is maximal either for $\lambda = 0$, i.e., for the single exposure to dose D, or for $\lambda = \frac{1}{4}$, i.e., for $D_1 = D_2 = D/2$. In the latter case,

$$\Phi_2^*(1; D) = \Phi\left(\frac{D}{2}, \frac{D}{2}\right) = \Phi(D) + \frac{1}{4} I(D) \quad .$$

Therefore, we can formulate the following proposition for $m = 1$.

3.2.1 Proposition 1

$$\Phi_2^*(1; D) = \max\left\{\Phi(D), \Phi\left(\frac{D}{2}, \frac{D}{2}\right)\right\}, \quad D > 0 \quad . \qquad (2.4)$$

Remark 1 — The equality 2.4 is true for every finite measure μ on R_+.

Remark 2 — Suppose set $\Delta_{1,1}^*$ of optimal single doses contains a unique element D^*. Also let function $\Phi(D)$ increase for $0 \leq D \leq D^*$ and decrease for $D \geq D^*$. Then by 2.2, $I(D) \leq 0$ for $0 \leq D \leq D^*$ and $I(D) \geq 0$ for $D \geq D^*$. Consequently,

$$\Phi_2^*(1; D) = \begin{cases} \Phi(D), & 0 \leq D \leq D^* \\ \Phi\left(\dfrac{D}{2}, \dfrac{D}{2}\right), & D > D^* \end{cases} \quad .$$

Now turn to the general fractionation problem with unlimited number of fractions. The solution is based crucially on the following generalization of Proposition 1. In its formulation, we use the notation

$$\mathcal{D}_n(D) = \underbrace{\left(\frac{D}{n}, \frac{D}{n}, \ldots, \frac{D}{n}\right)}_{n} \quad .$$

3.2.2 Proposition 2

Let $m = 1$. For every total dose $D > 0$ and for each $N \in \mathbb{N}$, there is an integer $n \leq N$ such that

$$\Phi(\mathscr{D}_n(D)) = \Phi_N^*(1; D) \quad .$$

In other words, among optimal fractionation schemes with a bounded number of fractions, there always exists a uniform one.

Proof — Let n be the least number of nonzero components for schemes in $\Delta_N^*(1; D)$, and let Δ^* be the set of schemes in $\Delta_N^*(1; D)$ with exactly n nonzero components. Clearly, the set Δ^* is nonempty. The case $n = 1$ is trivial. Suppose $n \geq 2$ and show that $\mathscr{D}_n(D) \in \Delta^*$.

On the contrary, let $\mathscr{D}_n(D) \notin \Delta^*$. Then,

$$\delta_0 : = \inf_{\mathscr{D} \in \Delta^*} \delta(\mathscr{D}) > 0 \quad ,$$

where $\delta(\mathscr{D})$ is the difference between the greatest and the smallest fractions in a scheme $\mathscr{D}$, and there is a scheme $\mathscr{D}_0 \in \Delta^*$ such that $\delta(\mathscr{D}_0) = \delta_0$. Rearranging the components of $\mathscr{D}_0$ in nondecreasing order, we write the scheme $\mathscr{D}_0$ in the form

$$\mathscr{D}_0 = (\underbrace{D_{max}, \ldots, D_{max}}_{i}, D_{i+1}, \ldots, D_{i+k}, \underbrace{D_{min}, \ldots, D_{min}}_{j})$$

where D_{max} and $D_{min} > 0$ are the maximal and minimal components of $\mathscr{D}_0$; $i, j \geq 1$; $k \geq 0$; $i + j + k = n$. Suppose for instance that $i \geq j$. Set $D' = D_{min} + D_{max}$, and define the measure

$$d\nu(x) = e^{-x(D-D')}(1 + D_{max})^{i-1}(1 + xD_{min})^{j-1} \prod_{r=i+1}^{i+k} (1 + xD_r)d\mu(x) \quad .$$

Applying Proposition 1 and Remark 1 after it, and using the definition of the number n, we see that the scheme

$$\mathscr{D}_1 = \left(\underbrace{D_{max}, \ldots, D_{max}}_{i-1}, \frac{D'}{2}, D_{i+1}, \ldots, D_{i+k}, \frac{D'}{2}, \underbrace{D_{min}, \ldots, D_{min}}_{j-1}\right)$$

belongs to Δ^*. Repeating this argument $j - 1$ times more, we get the scheme

$$\mathcal{D}_2 = \Big(\underbrace{D_{max}, \ldots, D_{max}}_{i-j}, \underbrace{\frac{D'}{2}, \ldots, \frac{D'}{2}}_{j}, D_{i+1}, \ldots, D_{i+k}, \underbrace{\frac{D'}{2}, \ldots, \frac{D'}{2}}_{j}\Big)$$

in Δ^* such that $\delta(\mathcal{D}_2) < \delta(\mathcal{D}_0) = \delta_0$.

This contradiction leads us to the conclusion that $\mathcal{D}_n(D) \in \Delta^*$, which completes the proof.

Corollary —

$$\Phi^*(1; D) = \sup_{n \in \mathbb{N}} \Phi(\mathcal{D}_n(D)) = \sup_{n \in \mathbb{N}} \int_0^\infty e^{-xD}\Big(1 + \frac{xD}{n}\Big)^n d\mu(x) \quad . \quad (2.5)$$

Since the integral in the right-hand side of 2.5 tends to zero as $n \to \infty$, the supremum in 2.5 is attained for some finite number n^* depending on D. Thus, for $m = 1$, the optimal fractionation scheme can always be found among uniform ones. This property holds for value $m = 1$ only, and it can be used in practice for distinguishing this case from the others.

An estimate of n^* can be extracted from the inequality presented below. For the sake of our further considerations, we formulate it in a more general form than it is needed just now. The positive and the negative variations of a measure μ are denoted by μ_+ and μ_-, respectively.

Lemma 2 — For every scheme $\mathcal{D} = (D_1, \ldots, D_n)$,

$$\Phi(\mathcal{D}) \le \frac{\sum\limits_{i=1}^n D_i^{m+1}}{(m+1)!} \int_0^\infty x^{m+1} d\mu_-(x), \quad m \in \mathbb{Z}_+ \quad . \quad (2.6)$$

Proof — We begin with the inequality

$$t - \ln\Big(1 + t + \ldots + \frac{t^m}{m!}\Big) \le \frac{t^{m+1}}{(m+1)!}, \quad t \in \mathbb{R}_+, \, m \in \mathbb{Z}_+ \quad ,$$

which can be easily checked by comparing the derivatives of its sides. Hence,

$$e^{-\frac{t^{m+1}}{(m+1)!}} \le e^{-t}\Big(1 + t + \ldots + \frac{t^m}{m!}\Big) \le 1, \, t \in \mathbb{R}_+, \, m \in \mathbb{Z}_+ \quad .$$

Further,

$$\Phi(\mathcal{D}) = \int_0^\infty \prod_{i=1}^n f_m(xD_i)d\mu_+(x) - \int_0^\infty \prod_{i=1}^n f_m(xD_i)d\mu_-(x) \le$$

$$\mu_+(R_+) - \int_0^\infty \prod_{i=1}^n e^{\frac{-(xD_i)^{m+1}}{(m+1)!}} d\mu_-(x) =$$

$$\mu_+(R_+) - \int_0^\infty e^{-\frac{x^{m+1}}{(m+1)!}\sum_{i=1}^n D_i^{m+1}} d\mu_-(x)$$

Since $e^{-t} \ge 1 - t$ for $t \ge 0$ and $\mu_+(R_+) = \mu_-(R_+)$, the previous estimate can be continued, leading directly to 2.6.

Corollary — For a uniform scheme, we have

$$\Phi(\mathcal{D}_n(D)) \le \frac{D^{m+1}}{(m+1)!n^m} \int_0^\infty x^{m+1}d\mu_-(x) \quad .$$

Practically, this formula can be used in the following way. Suppose $\overline{\Phi}$ is the greatest efficiency we have already calculated. Then, for further optimization, we have only to look for such uniform irradiation schemes $\mathcal{D}_n(D)$ that satisfy the condition $\Phi(\mathcal{D}_n(D)) \ge \overline{\Phi}$, i.e., with

$$n \le \left(\frac{D^{m+1}}{(m+1)!\overline{\Phi}} \int_0^\infty x^{m+1}d\mu_-(x) \right)^{1/m} \quad .$$

Note that every new value $\overline{\Phi}$ greater than previous ones diminishes the remaining run for n.

In the particular case $m = 1$, $\Phi(\mathcal{D}_n.(D)) \ge \Phi(D)$. Thus, we obtain the following theorem.

3.2.3 Theorem 3

Suppose that $\int_0^\infty x^2d\mu_-(x) < \infty$. Then,

$$\Phi^*(1; D) = \max_{1 \le n \le N} \int_0^\infty e^{-xD} \left(1 + \frac{xD}{n} \right)^n d\mu(x) \quad ,$$

where N is the integer part of $\frac{D^2}{2\Phi(D)} \int_0^\infty x^2d\mu_-(x)$.

Table 8

Results of Solution of the Optimal Fractionation Problem in the Case m = 1

Variant No.	$D = D_{opt}$	$\Phi(D_{opt})$	$\Phi^*(1; D)$	n^*	N
1	7	0.341	0.342	2	11
2	8	0.359	0.364	2	12
3	10	0.069	0.071	2	38
4	6	0.284	0.284	1	11
5	1.4	0.006	0.007	2	13
6	8	0.689	0.689	1	12

In the numerical example, we used the optimal single dose D_{opt} as the total dose D. The results of the solution of the optimal fractionation problem in the case m = 1 are presented in Table 8.

Let m = 2. In this case, the solution of the dichotomy problem is elementary, and the results have a simple geometrical interpretation. That is the reason why the case m = 2 is considered separately.

We have

$$\left(1 + xD_1 + \frac{(xD_1)^2}{2}\right)\left(1 + xD_2 + \frac{(xD_2)^2}{2}\right) =$$

$$1 + xD + \frac{(xD)^2}{2} + x^3\frac{DD_1D_2}{2} + x^4\frac{(D_1D_2)^2}{4} \quad .$$

Hence, for $\mathscr{D} = (D_1, D_2)$,

$$\Phi(\mathscr{D}) = \Phi(D) + \frac{\lambda}{2}I_1(D) + \frac{\lambda^2}{4}I_2(D) \quad ,$$

where λ is defined as above and

$$I_1(D) = \int_0^\infty e^{-xD}(xD)^3 d\mu(x), \quad I_2(D) = \int_0^\infty e^{-xD}(xD)^4 d\mu(x) \quad .$$

Maximization of function $\lambda \mapsto \frac{\lambda}{2}I_1(D) + \frac{\lambda^2}{4}I_2(D)$, $\lambda \in \left[0, \frac{1}{4}\right]$, yields the results shown in Table 9 and presented graphically in Figure 10. Note that when μ runs over the set of all differences of probability measures on R_+, the pairs $(I_1(D), I_2(D))$ for fixed D fill some convex symmetric set in R^2 containing zero and having nonvoid interior (see Figure 10).

Now consider the case of arbitrary m ≥ 2. Set

Table 9

Results of Solution of the Optimal Dichotomy Problem

Variant No.	D	Φ(D)	D_1^*	D_2^*	$\Phi_2^*(2; D)$	Φ(D)	D_1^*	D_2^*	$\Phi_2^*(3; D)$
			m = 2					**m = 3**	
1	50	0.014	25	25	0.039	0.031	25	25	0.117
2	60	0.038	30	30	0.078	0.070	30	30	0.158
3	50	0.009	25	25	0.020	0.017	25	25	0.043
4	40	0.016	20	20	0.057	0.045	20	20	0.156
5	4.7	0.001	2.35	2.35	0.017	0.021	4.7	0	0.021
6	80	0.033	40	40	0.095	0.079	40	40	0.239

Variant No.	D	Φ(D)	D_1^*	D_2^*	$\Phi_2^*(5; D)$	Φ(D)	D_1^*	D_2^*	$\Phi_2^*(10; D)$
			m = 5					**m = 10**	
1	50	0.120	25	25	0.385	0.525	38.6	11.4	0.618
2	60	0.165	30	30	0.359	0.457	37.9	22.1	0.583
3	50	0.045	25	25	0.106	0.134	43.5	6.5	0.142
4	40	0.163	20	20	0.392	0.478	36.6	3.4	0.486
5	4.7	0.016	4.7	0	0.016	0.0004	4.7	0	0.0004
6	80	0.249	40	40	0.614	0.779	40	40	0.969

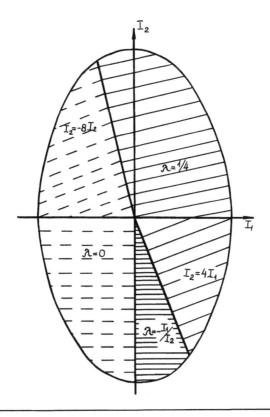

Figure 10
See text for explanations. (From Hanin, L. G., Rachev, S. T., and Yakovlev, A. Yu., *Appl. Probab.*, 25, 1, 1993. With permission.)

$$P_m(t) = \sum_{k=0}^{m} \frac{t^k}{k!} \quad ,$$

and note that

$$P_m(\lambda_1 t)P_m(\lambda_2 t) = \sum_{k=0}^{2m} \frac{t^k}{k!} S_{k,m}(\lambda_1, \lambda_2) \quad ,$$

where

$$S_{k,m}(\lambda_1, \lambda_2) = \sum_{\substack{i+j=k \\ 0 \le i, j \le m}} \binom{k}{i} \lambda_1^i \lambda_2^j \quad .$$

If $\lambda_1 + \lambda_2 = 1$, then for $0 \leq k \leq m$, we have

$$S_{k,m}(\lambda_1, \lambda_2) = \sum_{\substack{i+j=k \\ i,j \geq 0}} \binom{k}{i} \lambda_1^i \lambda_2^j = (\lambda_1 + \lambda_2)^k = 1 \quad .$$

Therefore, for scheme $\mathcal{D} = (D_1, D_2)$,

$$\Phi(\mathcal{D}) = \int_0^\infty e^{-xD} P_m(\lambda_1 xD) P_m(\lambda_2 xD) d\mu(x) =$$

$$\int_0^\infty e^{-xD} P_m(xD) d\mu(x) + \sum_{k=m+1}^{2m} \frac{M_k(D)}{k!} S_{k,m}(\lambda_1, \lambda_2) \quad ,$$

where $\lambda_1 = D_1/D$, $\lambda_2 = D_2/D$, as above, and

$$M_k(D) = \int_0^\infty e^{-xD}(xD)^k d\mu(x) \quad . \tag{2.7}$$

The polynomial $S_{k,m}(\lambda_1, \lambda_2)$ is a symmetric one. Hence, it can be represented as a polynomial of two elementary symmetric functions $\lambda_1 + \lambda_2$ and $\lambda_1 \lambda_2$. Thus, for $\lambda_1 + \lambda_2 = 1$, $S_{k,m}(\lambda_1, \lambda_2)$ becomes a polynomial in one variable $\lambda = \lambda_1 \lambda_2$, and we aim to find its explicit expression for $m + 1 \leq k \leq 2m$.
    For odd k,

$$S_{k,m}(\lambda_1, \lambda_2) = \sum_{\substack{i+j=k \\ 0 \leq i < j \leq m}} \binom{k}{i} (\lambda_1 \lambda_2)^i (\lambda_1^{j-i} + \lambda_2^{j-i}) =$$

$$\sum_{i=k-m}^{[k/2]} \binom{k}{i} \lambda^i (\lambda_1^{k-2i} + \lambda_2^{k-2i}) \quad , \tag{2.8}$$

where [ ] denotes the integer part. Similarly, for even k,

$$S_{k,m}(\lambda_1, \lambda_2) = \sum_{i=k-m}^{(k/2)-1} \binom{k}{i} \lambda^i (\lambda_1^{k-2i} + \lambda_2^{k-2i}) + \binom{k}{\frac{k}{2}} \lambda^{k/2} \quad . \tag{2.9}$$

Suppose for instance that $\lambda_1 \leq \lambda_2$. The equalities $\lambda_1 + \lambda_2 = 1$, $\lambda_1 \lambda_2 = \lambda$ imply

$$\lambda_1 = \frac{1}{2}(1 - \sqrt{1 - 4\lambda}), \quad \lambda_2 = \frac{1}{2}(1 + \sqrt{1 - 4\lambda}) \quad .$$

Therefore, for all $p \in \mathbb{N}$,

$$\lambda_1^p + \lambda_2^p = 2^{-p}\left(\sum_{u=0}^{p}(-1)^u \binom{p}{u}(1 - 4\lambda)^{u/2} + \sum_{u=0}^{p}\binom{p}{u}(1 - 4\lambda)^{u/2}\right) =$$

$$2^{-p+1}\sum_{v=0}^{[p/2]}\binom{p}{2v}(1 - 4\lambda)^v = 2^{-p+1}\sum_{v=0}^{[p/2]}\binom{p}{2v}\sum_{s=0}^{v}\binom{v}{s}(-4\lambda)^s =$$

$$2^{-p+1}\sum_{s=0}^{[p/2]}(-4\lambda)^s\sum_{j=0}^{[p/2]-s}\binom{p}{2s + 2j}\binom{s + j}{s} \quad .$$

Inserting this expression with $p = k - 2i$ into 2.8 and setting $r = i + s$, we find that for odd k, $m + 1 \le k \le 2m$,

$$S_{k,m}(\lambda_1, \lambda_2) = \sum_{r=k-m}^{[k/2]}\alpha_{k,r}\lambda^r \quad , \tag{2.10}$$

where

$$\alpha_{k,r} = (-1)^r 2^{-k+2r+1}\sum_{i=k-m}^{r}(-1)^i\binom{k}{i}\gamma_{k,r,i} \, , \quad k - m \le r \le \left[\frac{k}{2}\right] \quad , \tag{2.11}$$

and

$$\gamma_{k,r,i} = \sum_{j=0}^{[k/2]-r}\binom{k - 2i}{2(r - i + j)}\binom{r - i + j}{r - i}, \quad 1 \le i \le r \le \left[\frac{k}{2}\right] \quad , \tag{2.12}$$

$$\left(\text{with } \binom{0}{0}: = 1\right).$$

Define also for even k

$$\gamma_{k,\frac{k}{2},\frac{k}{2}}: = \frac{1}{2} \quad . \tag{2.13}$$

Then, as it follows from 2.9, Formulas 2.10 and 2.11 are valid for all k.

We have finally

$$\Phi(\mathcal{D}) = \Phi(\mathcal{D}) + \sum_{r=m+1}^{2m}\frac{M_k(D)}{k!}\sum_{r=k-m}^{[k/2]}\alpha_{k,r}\lambda^r = \Phi(D) + \sum_{r=1}^{m}\beta_{m,r}\lambda^r$$

with

$$\beta_{m,r} = \sum_{k=\max(2r,m+1)}^{m+r} \frac{M_k(D)}{k!} \alpha_{k,r} \, , \; 1 \leq r \leq m \; . \qquad (2.14)$$

So, the dichotomy problem is reduced to maximizing the polynomial

$$Q_m(\lambda) = \sum_{r=1}^{m} \beta_{m,r} \lambda^r$$

with the coefficients defined by means of 2.14, 2.11, 2.12, 2.13, and 2.7, on the interval $\left[0, \dfrac{1}{4}\right]$. This problem can be solved by standard algebraic methods to any desired precision.

Besides the case $m = 1$, numerical solution of the dichotomy problem was performed for $m = 2, 3, 5, 10$. For each pair of radiosensitivity distributions, the total dose D was chosen to be the same for every such m. In all cases, the optimal fractionation schemes (D_1^*, D_2^*) and the corresponding optimal efficiencies $\Phi_2^*(m; D)$ were computed (see Table 9).

Unlike the case $N = 2$, precise methods of the optimal fractionation of the total dose without limitation of the number of fractions are not elaborated so far. Nevertheless, the following iterative process enables us to increase substantially the optimal two-fraction irradiation efficiency.

As an initial approximation, we take the single exposure scheme, $\mathcal{D}^{(1)} := D$. Let $\mathcal{D}^{(2)} = (D_1^{(2)}, D_2^{(2)})$, $D_1^{(2)} + D_2^{(2)} = D$, be the optimal fractionation scheme obtained by means of the above described dichotomy procedure. Applying the latter to the dose $D_1^{(2)}$ and to the measure

$$d\nu(x) = e^{-xD_2^{(2)}} P_m(xD_2^{(2)}) d\mu(x)$$

we get the partition $D_1^{(2)} = D_1' + D_1''$. Repeating the procedure for the dose $D_2^{(2)}$, we obtain the partition $D_2^{(2)} = D_2' + D_2''$. Comparing the schemes $(D_1', D_1'', D_2^{(2)})$ and $(D_1^{(2)}, D_2', D_2'')$, we choose the one with greater efficiency and denote it $\mathcal{D}^{(3)} = (D_1^{(3)}, D_2^{(3)}, D_3^{(3)})$. Suppose we have already found the scheme $\mathcal{D}^{(n-1)} = (D_1^{(n-1)}, \ldots, D_{n-1}^{(n-1)})$. Applying the dichotomy process to each of its fractions (with the corresponding measures instead of μ) and selecting among $n - 1$ schemes thus obtaining the most efficient one, we get the scheme $\mathcal{D}^{(n)} = (D_1^{(n)}, \ldots, D_n^{(n)})$, and so on. The iteration of the dichotomy process is stopped, if either $\mathcal{D}^{(n)} = \mathcal{D}^{(n-1)}$ for some n or the efficiency increment $\Phi(\mathcal{D}^{(n)}) - \Phi(\mathcal{D}^{n-1})$ is small enough. Then the scheme $\mathcal{D}^{(n)}$ is the required diadic regimen.

Remark — In the case $\Delta = \Delta_N(D)$, the diadic process goes on for $n \le N$ unless it has not stopped earlier.

The properties of the diadic fractionation regimens were studied by means of a numerical experiment. The values of the parameter m chosen for calculations are m = 2, 5. The total dose D was taken from the interval $[\bar{D}_{opt}, 2\bar{D}_{opt}]$, where $\bar{D}_{opt}$ is the sum of two doses with the maximal efficiency among *all* two-fraction irradiation schemes.

The results of the diadic fractionation are given in Table 10 in comparison with the single and the optimal two-fraction efficiencies for the same total dose.

Analysis of the numerical results leads us to the following preliminary conclusions.

1. For $D \ge \bar{D}_{opt}$, the diadic partition provides a substantial gain (and even increasing with the growth of the total dose) in therapeutic efficiency as compared with the single exposure to the dose D and with the optimal two-fraction irradiation with the same total dose.
2. The number of fractions in the diadic partition of the dose D is nondecreasing with respect to D.
3. Unlimited crumbling of the total dose is not efficient (the optimal regimen in the case $\Delta = \Delta(D)$ is likely to be discrete).
4. In the case $\Delta = \Delta(D)$, the diadic scheme for $D \ge \bar{D}_{opt}$ has a tendency to approach a uniform one.

It is interesting to compare the above results with those obtained without any restrictions on the irradiation regimens. We treat this unconditional optimization problem numerically by applying the flexible simplex algorithm[22] to determine the optimal values of doses D_i, $D_i > 0$, i = 1, $\ldots$, n for every fixed value of n. In doing so, we have to confine ourselves to some small values of n. Accordingly, Figures 11 to 13 represent the optimal values of Φ only for n = 1, $\ldots$, 8. However, this does not prevent us from observing some interesting regularities. The numerical experiments were carried out by Kulagin. Figure 11 corresponds to variant 1 of the radiosensitivity distributions for normal and neoplastic cells (see Table 6 and Figure

Table 10

Results of the Diadic Fractionation Procedure Applied to Various Total Doses D

Variant No.	Total Dose D	Optimal Scheme	n	$\Phi_n^*(m; D)$	$\Phi_2^*(m; D)$	$\Phi(D)$
		m = 2				
2	17.5	(5.76; 8.75; 2.99)	3	0.441	0.436	0.352
	33	$D_i = 4.125$, $i = 1, \ldots, n$	8	0.476	0.253	0.147
3	19	(8.89; 9.50; 0.61)	3	0.094	0.093	0.071
	38	$D_i = 4.75$, $i = 1, \ldots, n$	8	0.107	0.039	0.019
4	17	$D_i = 4.25$, $i = 1, \ldots, n$	4	0.387	0.340	0.223
	26	$D_i = 3.25$, $i = 1, \ldots, n$	8	0.380	0.190	0.089
		m = 5				
2	40	(15.81; 18.79; 4.19; 1.21)	4	0.536	0.525	0.336
	60	$D_i = 15$, $i = 1, \ldots, n$	4	0.556	0.359	0.165
3	40	(18.39; 19.49; 1.61; 0.51)	4	0.127	0.126	0.072
	60	$D_i = 15$, $i = 1, \ldots, n$	4	0.136	0.078	0.028
4	35	(13.67; 16.23; 3.83; 1.27)	4	0.444	0.431	0.230
	45	$D_i = 11.25$, $i = 1, \ldots, n$	4	0.458	0.338	0.111

From Hanin, L. G., Rachev, S. T., and Yakovlev, A. Yu., *Appl. Probab.*, 25, 1, 1993. With permission.

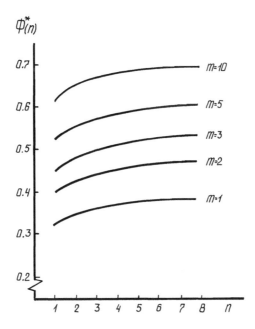

Figure 11
See text for explanations.

9), while Figures 12 and 13 correspond to variants 3 and 5. Shown in Figures 14 to 16 are the total dose values for which the maximum Φ^* of the functional Φ is attained. For all variants, the optimal schemes are close to uniform ones, and the value Φ^* increases with increasing n. The tendency to saturation manifests itself when the dependence of the optimal value Φ^* on the number of fractions, n, is under consideration (Figures 11 to 13), while the dependence on n of the optimal total dose seems to be approximately linear (Figures 14 to 16). For all variants tested in these experiments, the optimal values of both the functional and the total dose increase with the growth of the critical number of hits m. These results are worth comparing with those presented in Chapter 2.

Another optimal control problem, which seems to be important from the biomedical point of view, implies restriction imposed on the expected survival of normal cells. This problem was formulated in Section 2.2 as the search for a fractionated irradiation scheme for which the maximal value of the functional $\Phi(\mathfrak{D})$ on the set

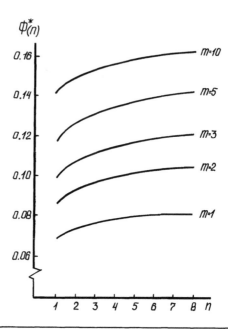

Figure 12
See text for explanations.

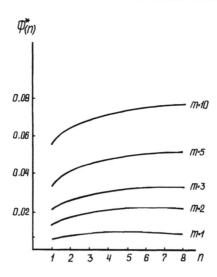

Figure 13
See text for explanations.

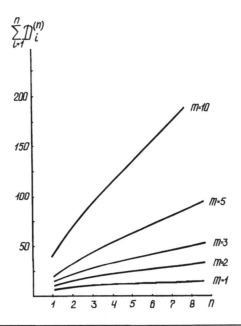

Figure 14
See text for explanations.

$$\Delta^{\gamma} := \left\{ \mathcal{D} \in \Delta_0 : \Phi_0(\mathcal{D}) = \int_0^{\infty} g(x, \mathcal{D})d\mu_0(x) \geq \gamma \right\}, \quad 0 < \gamma < 1 \quad ,$$

is attained, where γ is the prescribed lowest level of the expected survival of normal cells.

Without any loss of generality, one may consider only the values of γ satisfying the following condition: $\gamma > \gamma^* = \Phi_0(\mathcal{D}^*)$, where $\mathcal{D}^*$ is the unconditional (see above) optimal irradiation scheme. Moreover, leaving behind trivial cases, the inequality type restriction $\Phi_0(\mathcal{D}) \geq \gamma$ may be replaced by the equality $\Phi_0(\mathcal{D}) = \gamma$. It is obvious that both the optimal total dose and the optimal efficiency decrease with increasing γ, and this is confirmed by numerical computations (Figure 17). Besides, as it follows from numerical experiments carried out for $n = 1, \ldots, 8$, the optimal fractionation is uniform in this case, i.e., $D_1^* = D_2^* = \ldots = D_n^*$. In another set of numerical experiments, the optimal dichotomy problem was solved for dose D equal to the total dose for the optimal regimen that had been obtained for $n = 2$ and for a given value of γ. Only slight differences were observed between these two optimal (but in different senses) irradiation schemes for all

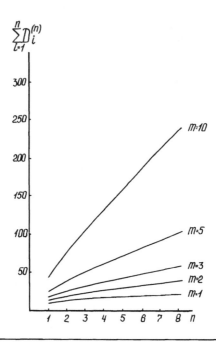

Figure 15
See text for explanations.

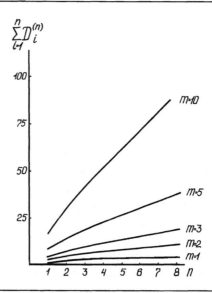

Figure 16
See text for explanations.

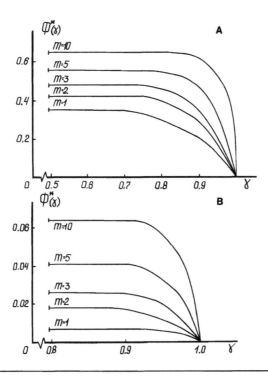

Figure 17
The optimal value of the efficiency functional Φ as a function of γ. Numerical experiments were conducted for $n = 2$ and for variants 1 and 5 of the function $F_0 - F_1$. (A) variant 1, (B) variant 5.

combinations of the radiosensitivity parameters involved in those experiments.

3.3 STEPWISE CONSTRUCTION OF THE OPTIMAL IRRADIATION SCHEME

In this section, we construct an optimal, in some natural sense, irradiation scheme by means of a method quite different from those used before. The main idea is to break the optimization process into several steps — each of them specifying an optimal single dose for cell populations survived on previous steps.

Suppose normal and neoplastic tissues with radiosensitivity distributions μ_0 and μ_1 are exposed to a dose D_1. Then the radiosensitivity distributions for survived cell populations in both tissues are

$$d\mu_i^{(1)} : = \frac{g(x, D_1)d\mu_i(x)}{\displaystyle\int_0^\infty g(x, D_1)d\mu_i(x)}, \qquad i = 0, 1 \quad.$$

Subsequent application of dose D_2 provides the efficiency

$$\Psi(D_1, D_2) : = \int_0^\infty g(x, D_2)d(\mu_0^{(1)} - \mu_1^{(1)})(x) =$$

$$\frac{\displaystyle\int_0^\infty g(x, D_1)g(x, D_2)d\mu_0(x)}{\displaystyle\int_0^\infty g(x, D_1)d\mu_0(x)} - \frac{\displaystyle\int_0^\infty g(x, D_1)g(x, D_2)d\mu_1(x)}{\displaystyle\int_0^\infty g(x, D_1)d\mu_1(x)}$$

while the new radiosensitivity distributions for survived cells become

$$d\mu_i^{(2)} : = \frac{g(x, D_2)d\mu_i^{(1)}(x)}{\displaystyle\int_0^\infty g(x, D_2)d\mu_i^{(1)}(x)} = \frac{g(x, D_1)g(x, D_2)d\mu_i(x)}{\displaystyle\int_0^\infty g(x, D_1)g(x, D_2)d\mu_i(x)}, \ i = 0, 1 \quad.$$

Similarly, when a sequence of doses $\mathscr{D} = (D_1, \ldots, D_n)$ is applied, the radiosensitivity distributions for survived cell populations are

$$d\mu_i^{(n)} : = \frac{g(x, D_n)d\mu_i^{(n-1)}(x)}{\displaystyle\int_0^\infty g(x, D_n)d\mu_i^{(n-1)}(x)} = \frac{\displaystyle\prod_{j=1}^n g(x, D_j)d\mu_i(x)}{\displaystyle\int_0^\infty \prod_{j=1}^n (x, D_j)d\mu_i(x)}, \ i = 0, 1 \quad.$$

and the efficiency is equal to

$$\Psi(\mathscr{D}) = \int_0^\infty g(x, D_n)d(\mu_0^{(n-1)} - \mu_1^{(n-1)})(x) =$$

$$\frac{\displaystyle\int_0^\infty \prod_{j=1}^n g(x, D_j)d\mu_0(x)}{\displaystyle\int_0^\infty \prod_{j=1}^{n-1} g(x, D_j)d\mu_0(x)} - \frac{\displaystyle\int_0^\infty \prod_{j=1}^n g(x, D_j)d\mu_1(x)}{\displaystyle\int_0^\infty \prod_{j=1}^{n-1} g(x, D_j)d\mu_1(x)} \quad.$$

Functional $\Psi(\mathscr{D})$ can be rewritten in the form

$$\Psi(\mathscr{D}) = \int_0^\infty \tilde{g}_0(x, \mathscr{D})d\mu_0(x) - \int_0^\infty \tilde{g}_1(x, \mathscr{D})d\mu_1(x) \quad,$$

Table 11
Results of the Stepwise Optimization Procedure

Variant No.	$\hat{D}_1$	$\hat{D}_2$	$\tilde{D}_1$	$\tilde{D}_2$	D_1^*	D_2^*
	$\Phi(\hat{D}_1, \hat{D}_2)$		$\Phi(\tilde{D}_1, \tilde{D}_2)$		$\Phi(D_1^*, D_2^*)$	
1	10	12	11	11	7.5	7.5
	0.343		0.345		0.434	
2	12	14	13	13	8	8
	0.341		0.342		0.440	
3	14	13	13.5	13.5	9.5	9.5
	0.072		0.072		0.094	
4	9	11	10	10	6.5	6.5
	0.286		0.288		0.355	
5	3	2	2.5	2.5	2	2
	0.015		0.016		0.017	
6	12	14	13	13	8.5	8.5
	0.718		0.719		0.810	

where

$$\tilde{g}_i(x, \mathcal{D}) = \frac{\prod\limits_{j=1}^{n} g(x, D_j)}{\int_0^\infty \prod\limits_{j=1}^{n-1} g(x, D_j)d\mu_i(x)}, \quad i = 0, 1 \quad,$$

and are noncommutative response functions. Note that the value $\tilde{g}_i(x, \mathcal{D})$ is equal to the conditional survival probability for a cell with radiosensitivity x in the corresponding population to survive dose D_n provided that it has survived being exposed to doses $D_1, \ldots, D_{n-1}$.

To obtain the stepwise optimal irradiation scheme, we proceed in the following way. In the first step, we find the dose $\hat{D}_1$ such that $\Psi(D)$ is maximal. In the second step, we get the dose $\hat{D}_2$ for which $\Psi(D, \hat{D}_1)$ is maximal, and so on. The optimization process goes on while $\hat{\mathcal{D}}_n = (\hat{D}_1, \ldots, \hat{D}_n)$ belongs to the prescribed set Δ or, in the case when no restrictions on irradiation schemes are imposed, until the increment $\Phi(\hat{\mathcal{D}}_n) - \Phi(\hat{\mathcal{D}}_{n-1})$ becomes insignificant.

The procedure of stepwise optimization is much easier than those of simultaneous optimization of fractions. Its effectiveness is demonstrated by the numerical example in which calculations were carried out for m = 2 and for all variants of pairs of distributions μ_0, μ_1. Compared in Table 11 are:

1. Optimal unconditional two-fraction scheme (D_1^*, D_2^*) and the corresponding efficiency $\Phi^*(\Delta_2) = \Phi(D_1^*, D_2^*)$;

2. Optimal stepwise two-fraction scheme $(\hat{D}_1, \hat{D}_2)$ with the efficiency $\Phi(\hat{D}_1, \hat{D}_2)$;
3. Optimal partition $(\tilde{D}_1, \tilde{D}_2)$ of the total dose $D = \hat{D}_1 + \hat{D}_2$ and its efficiency $\Phi^*(\Delta_2(D)) = \Phi(\tilde{D}_1, \tilde{D}_2)$.

The data obtained show that, at least for the two-fraction regimens, the stepwise optimization process yields an irradiation scheme that is quite close to the optimal dichotomic one with the same total dose, and that the difference between the optimal stepwise and the unconditional efficiencies is not significant.

3.4 STABILITY PROPERTIES OF THE EFFICIENCY FUNCTIONAL

In general, the radiosensitivity distributions μ_0 and μ_1 are unobservable and can scarcely be prescribed on an empirical basis. What might be known about them is their estimates, i.e., some approximating distributions $\hat{\mu}_0$ and $\hat{\mu}_1$, obtained with the aid of expert knowledge. These distributions give rise to the functional

$$\hat{\Phi}(f) : = \int_0^\infty f(x)d(\hat{\mu}_0 - \hat{\mu}_1)(x) \quad .$$

Accordingly, instead of each true extreme value Φ^* of the functional Φ (i.e., the supremum of Φ over a certain set of response functions, for example, over W, L_M, or $\{g(\cdot, \mathcal{D}): \mathcal{D} \in \Delta\}$) we are in a position to calculate only the corresponding extremal value $\hat{\Phi}^*$ for functional $\hat{\Phi}$.

The problem arising here is whether the difference between $\hat{\Phi}^*$ and Φ^* is small whenever the distributions $\hat{\mu}_0$ and $\hat{\mu}_1$ are close, in a sense, to μ_0 and μ_1, respectively.

Denote by $\hat{F}_0$, $\hat{F}_1$, and $\hat{F}$ the distribution functions of measures $\hat{\mu}_0$, $\hat{\mu}_1$ and $\hat{\mu} = \hat{\mu}_0 - \hat{\mu}_1$. For every function $f \in W$, we have

$$\hat{\Phi}(f) - \Phi(f) = \int_0^\infty fd(\hat{\mu} - \mu) = \int_0^\infty (F - \hat{F})(x)f'(x)dx \leq$$

$$\sup_{x \in R_+} |(\hat{F} - F)(x)| = \xi_0(\hat{\mu}, \mu) \leq \xi_0(\hat{\mu}_0, \mu_0) + \xi_0(\hat{\mu}_1, \mu_1) \quad .$$

Similarly, for $f \in L_M$,

$$\hat{\Phi}(f) - \Phi(f) \leq \xi_{M-1}(\hat{\mu}, \mu) \leq \xi_{M-1}(\hat{\mu}_0, \mu_0) + \xi_{M-1}(\hat{\mu}_1, \mu_1)$$

(see Section 2.3). Note also that

$$\xi_{M-1}(\hat{\mu}, \mu) \leq M\rho(\hat{\mu}, \mu) \quad,$$

where

$$\rho(\hat{\mu}, \mu) := \int_0^\infty |(\hat{F} - F)(x)|dx \quad.$$

Thus, using the symmetry argument, we obtain

$$|\hat{\Phi}_w - \Phi_w| \leq \xi_0(\hat{\mu}_0, \mu_0) + \xi_0(\hat{\mu}_1, \mu_1) \quad,$$
$$|\hat{\Phi}_{L_M} - \Phi_{L_M}| \leq \xi_{M-1}(\hat{\mu}_0, \mu_0) + \xi_{M-1}(\hat{\mu}_1, \mu_1) \leq$$
$$M(\rho(\hat{\mu}_0, \mu_0) + \rho(\hat{\mu}_1, \mu_1)) \quad.$$

Also, for every set of admissible schemes Δ, we have the estimate

$$|\hat{\Phi}^*(\Delta) - \Phi^*(\Delta)| \leq \xi_0(\hat{\mu}_0, \mu_0) + \xi_0(\hat{\mu}_1, \mu_1) \quad.$$

In the case $\Delta \subset \Delta(D)$, it can be amplified by

$$|\hat{\Phi}^*(\Delta) - \Phi^*(\Delta)| \leq a_m D(\rho(\hat{\mu}_0, \mu_0) + \rho(\hat{\mu}_1, \mu_1)) \leq$$
$$\frac{D}{\sqrt{2\pi m}} (\rho(\hat{\mu}_0, \mu_0) + \rho(\hat{\mu}_1, \mu_1)) \quad,$$

and we may choose the best one among these two estimates.

The estimates obtained above reveal the stability of functional Φ, its extremal values, and its theoretical upper bounds with respect to perturbation of distributions μ_0 and μ_1.

3.5 TRANSIENT PROCESSES IN TISSUES EXPOSED TO FRACTIONATED IRRADIATION

3.5.1 Introduction

The main idea exploited in the overwhelming majority of modern approaches to optimization of multiple dose irradiation of tumors is that responses of the normal and neoplastic tissues are essentially different as far as their temporal organization is concerned. It is commonly believed that the time factor plays an extremely important role not only in radiotherapy but also in tumor treatment of any kind. Definitely, it seems very attractive to introduce this factor into the model considered in the previous chapter and thus to include the time intervals between the consecutive fractions into the set of parameters to be optimized. However, it is clear that such a generalization of the model itself and the corresponding formulations of the optimal control problem would lead to insurmountable analytical difficulties. This is the very reason why we turn to computer simulation techniques to provide a more realistic description of the fractionated irradiation of tumors. The aim of this simulation study is to make some preliminary assessment of possible deviations that may occur from the solutions obtained in previous sections, if transient processes in radiation damage formation and postirradiation damage repair are taken into account. As in the numerical experiments presented above, only qualitative inferences might be of value from the practical point of view. Besides, such inferences are based inevitably on quite a restricted set of simulation experiments. A general description of computer simulation approach to analysis of radiobiological phenomena can be found in Reference 90.

3.5.2   Notation

m = Critical number of lesions a cell can be bear without being killed

r = Total number of reparons (repairing units)

μ_0, μ_1 = Radiosensitivity distributions for normal and neoplastic tissues, respectively

ν_0, ν_1 = Radiosensitivity distributions of reparons in normal and neoplastic tissues, respectively

P_0, P_1 = Probabilities of misrepair of a lesion for normal and neoplastic tissues, respectively

t_1 = Time of lesion repair

π_0, π_1 = Distributions of the time of lesion repair for normal and neoplastic tissues, respectively

t_p = Time of repair of a reparon

k_p = Number of lesions of a reparon

l_d = Number of unrepaired lesions beyond reparons

l_r = Number of lesions in a state of repair

l_w = Number of misrepaired lesions

r_f = Number of free reparons
r_l = Number of reparons repairing lesions
r_w = Number of misrepaired reparons
r_r = Number of reparons repairing reparons
r_d = Number of damaged reparons not in a state of repair
N = Total number of surviving cells in the simulated population
τ_j = Time interval between jth and (j + 1)th irradiations.

3.5.3 Relations Between the Model Parameters

$$r_l = l_r \quad ,$$
$$r_f + r_d + r_l + 2r_r + r_w = r \quad .$$

3.5.4 Basic Assumptions

1. Values m and r are nonrandom and the same for normal and neoplastic cells.

2. Death of a cell occurs, if and only if the number of irreparable lesions is greater than m, i.e., a cell dies when $l_w > m$, or if $r_f = r_l = r_r = 0$, when $l_w + l_d > m$.

3. A lesion (in a reparon or elsewhere in the genome of a cell) is thought of as a very local event. Thus, every lesion resulting from exposure to radiation cannot be the same as an existing one.

4. A reparon is damaged, if it has at least one lesion.

5. The time of repair of several lesions by the same reparon is the sum of repair times for all these lesions — the distribution for each of the latter times being equal to π.

6. A free reparon may repair either a lesion or a damaged reparon. The probability that a given reparon is assigned to repair of a lesion is

$$q_e = \frac{l_d}{\sum_{\rho=1}^{r} k_\rho + l_d}$$

while the probability that it is assigned to repair of a reparon ρ is

$$q_\rho = \frac{k_\rho}{\sum_{i=1}^{r} k_i + l_d} \quad .$$

7. Misrepair of a lesion or reparon is unrepairable.
8. A reparon is misrepaired if at least one of its lesions is misrepaired. If k lesions of the same reparon are repaired, then the probability of (at least one) misrepair is $1 - (1 - P)^k$.
9. Misrepair of a lesion or reparon can be established only after the repair process is completed.
10. If a reparon is damaged while performing repair of a lesion, then the repair process stops and the lesion remains unrepaired.
11. If a reparon is damaged while performing repair of a reparon, then the repair process stops, and all lesions in which repair has not been completed up to this moment remain unrepaired.
12. Damage which occurs to a reparon r_1 in a state of repair by a reparon r_2 will be repaired by the reparon r_2 after more recent lesions of the reparon r_1.

3.5.5 Computer Simulations

In order to conduct simulation experiments, it is necessary to specify the following characteristics of the model: m, r, μ_0, μ_1, ν_0, ν_1, P_0, P_1, π_0, π_1. In our simulations, the distributions μ_0, μ_1, ν_0, ν_1, π_0, π_1 were assumed to be gamma distributions. They were specified by the mean value M and the variation coefficient V — these parameters being the same for the pairs of distributions (μ_0, μ_1) and (ν_0, ν_1) in the set of simulation experiments that follows. Each of the cell populations was initialized to contain N = 500 cells, and all events occurring after irradiation were simulated for every individual cell. Every such simulation run was repeated L times. Pseudorandom numbers were generated by means of the General Purposes Simulation System (GPSS) (see Reference 90). Shown in Table 12 are particular values of the model parameters used in our simulations.

Optimal dynamic regimens of fractionated irradiation were sought with the aid of the simulation model, given the total dose D and the

Table 12
Values of Parameters Used in the Simulation Model

Variant No.	P_0	P_1	μ_0 M	μ_0 V	μ_1 M	μ_1 V	π_0 M	π_0 V	π_1 M	π_1 V
1	0.01	0.10	0.2	0.5	0.4	0.25	6.0	0.25	4.0	0.25
2	0.01	0.25								

number of dose fractions n. The following sets of admissible regimens were considered.

1. Uniform distribution of dose per fraction and equal time intervals between successive fractions,

$$\Delta(\tau) : = \left\{ D_i = \frac{D}{n}, i = 1, \ldots, n; \tau_j = \tau = \text{const}, j = 1, \ldots, n - 1 \right\} .$$

2. Equal time intervals between fractions,

$$\Delta(\mathcal{D}, \tau) : =$$

$$\left\{ \mathcal{D} = (D_1, \ldots, D_n), \sum_{i=1}^{n} D_i = D, \tau_j = \tau = \text{const}, j = 1, \ldots, n - 1 \right\} .$$

3. Arbitrary distributions of fractional doses and time intervals between them,

$$\Delta(\mathcal{D}, T) : = \left\{ \mathcal{D} = (D_1, \ldots, D_n), \sum_{i=1}^{n} D_i = D; T = (\tau_1, \ldots, \tau_{n-1}) \right\} .$$

Corresponding to these sets are the following optimal values of the tumor treatment efficiency:

$$\Phi^*_{\Delta(\tau)} = \Phi(\tau^*) ,$$
$$\Phi^*_{\Delta(\mathcal{D},\tau)} = \Phi(D^*_1, \ldots, D^*_n; \tau^*) ,$$
$$\Phi^*_{\Delta(\mathcal{D},T)} = \Phi(D^*_1, \ldots, D^*_n; \tau^*_1, \ldots, \tau^*_{n-1}) .$$

The set-up of simulation experiments implied obtaining all three optimal dynamic regimes for $D = 23$, $m = 2$, $k_p = 1$, $r = 10$, $n = 4$, $L = 30$, and the other parameters being presented in Table 12. In the simplest case of set $\Delta(\tau)$, the optimal interval, τ^*, between the successive fractions is determined from the plots of Φ vs. τ given in Figure 18. For the sets $\Delta(\mathcal{D}, \tau)$ and $\Delta(\mathcal{D}, T)$, the optimal regimes have been computed numerically by the flexible simplex method.[29] The results of these computations are presented in Tables 13 and 14 for the sets $\Delta(\mathcal{D}, \tau)$ and $\Delta(\mathcal{D}, T)$, respectively.

Scrutinizing the results obtained in this series of simulation experiments, one can formulate the following conclusions.

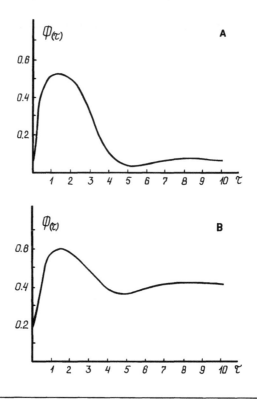

Figure 18
The plot of the efficiency functional Φ vs. τ for the set of admissible regimens $\Delta(\tau)$. (A) variant 1, (B) variant 2 (see Table 6).

Table 13

Optimal Dynamic Regimens of Fractionated Irradiation for the Sets $\Delta(\tau)$ and $\Delta(\mathcal{D}, \tau)$

Variant No.	$\Delta(\tau)$		$\Delta(\mathcal{D}, \tau)$		
	τ^*	Φ^*	$\mathcal{D}^*$	τ^*	Φ^*
1	1.00	0.518	5.89, 5.48, 6.09, 5.56	0.53	0.549
2	1.50	0.589	5.65, 6.05, 5.58, 5.72	1.10	0.620

Table 14

**Optimal Dynamic Regimens of Fractionated
Irradiation for the Set $\Delta(\mathcal{D}, T)$**

Variant No.	$\mathcal{D}^*$	T^*	Φ^*
1	5.94, 5.70, 5.55, 5.96	3.02, 1.73, 1.61	0.580
2	5.59, 5.46, 5.71, 6.30	2.17, 0.72, 1.80	0.650

1. When the optimal dynamic regimen for the set $\Delta(\tau)$ is constructed, the gain in efficiency appears to be greater for smaller differences between normal and neoplastic cells with respect to the value of misrepair probability (see Figure 18 and Table 13). Thus, in this case, the contribution of the time factor to optimization of fractionated irradiation is expected to be more significant for those tumors that are relatively radioresistant.

2. For the sets $\Delta(\mathcal{D}, \tau)$ and $\Delta(\mathcal{D}, T)$, the optimal dynamic regimen retains a tendency to be uniform with respect to the dose per fraction value.

3. The optimal regimen constructed for the set $\Delta(\mathcal{D}, T)$ is not uniform with respect to the time intervals between fractions. However, optimization of these intervals leads to slight increase in the value of the efficiency functional.

Conclusion

In conclusion, we summarize basic results described in this book.

The tumor irradiation efficiency Φ can be measured by the difference between the expected survival probabilities for normal and neoplastic cells, i.e.,

$$\Phi(g; \mathcal{D}) = \int_0^\infty g(x, \mathcal{D})d(\mu_0 - \mu_1)(x) \quad ,$$

where $g(x, \mathcal{D})$ is the survival probability of a cell with radiosensitivity x exposed to the sequence of doses $\mathcal{D} = (D_1, \ldots, D_n)$, and μ_0, μ_1 are radiosensitivity distributions for normal and neoplastic cells, respectively. In the special case of multihit-one target model,

$$g(x, \mathcal{D}) = e^{-x\sum_{i=1}^{n} D_i} \prod_{i=0}^{n} \sum_{k=0}^{m} \frac{(xD_i)^k}{k!} \quad , \tag{1}$$

where m is the maximal quantity of hits (lesions) the cells can bear without being killed.

The following problems have been formulated.

1. PROBLEM 1

To find the precise upper bound of the functional Φ over certain classes $\mathcal{F}$ of admissible response functions $g(x, \mathcal{D})$.

2. PROBLEM 2

To construct for model 1 an optimal irradiation scheme $\mathscr{D}^*$ within a prescribed set Δ of admissible irradiation schemes, such that $\Phi^*: = \sup_{\mathscr{D} \in \Delta} \Phi(g; \mathscr{D}) = \Phi(g; \mathscr{D}^*)$, and to obtain the corresponding optimal efficiency Φ^*.

These problems can be solved in the following cases:

1. Irradiation schemes $\mathscr{D} = (D_1, \ldots, D_n)$ are arbitrary;
2. $D_1 + \ldots + D_n \leq D$ for a given total dose D;
3. $D_1 + \ldots + D_n = D$ for a certain total dose D;
4. $\Phi_0(\mathscr{D}) \geq \gamma$, where $\Phi_0(\mathscr{D}) = \int_0^\infty g(x, \mathscr{D})d\mu_0(x)$ is the mean survival probability for normal cells and γ is a given survival level.

We may suppose without loss of generality that $\gamma > \gamma^* = \Phi_0(\mathscr{D}^*)$, where $\mathscr{D}^*$ is the unconditional optimal irradiation scheme in Case 1 (if $\gamma \leq \gamma^*$, then Cases 1 and 4 are equivalent).

In Case 1, there are weighty reasons to take for $\mathscr{F}$ the class W of all nonincreasing absolutely continuous functions f on $\mathbb{R}_+$ such that $f(0) = 1$ and $\lim_{x \to +\infty} f(x) = 0$. The solution of Problem 1 in this case is given by the following theorem.

3. THEOREM 1

$\sup_{f \in W} \Phi(f) = \sup_{x \in \mathbb{R}_+} (F_0 - F_1)(x)$, where F_i is the distribution function for the measure μ_i, $i = 0, 1$.

Denote

$$E = \left\{ t \in \mathbb{R}_+ : (F_0 - F_1)(t) = \sup_{x \in \mathbb{R}_+} (F_0 - F_1)(x) \right\} .$$

If (and only if) mesE > 0, then $\sup \Phi(f)$ over $f \in W$ is attained, and the optimal function $f^* \in W$ is given by the formula

$$f^*(x) = 1 - (\text{mesE})^{-1} \int_0^x \chi_E(t)dt , \qquad (2)$$

χ_E being the characteristic function of the set E.

In Case 2, $\mathcal{F}$ can be identified with the class L_M constituted by all nonincreasing functions f on $\mathbb{R}_+$ satisfying the Lipschitz condition $|f(x) - f(y)| \le M|x - y|$ with the constant $M = De^{-m}m^m/m!$, and such that $f(0) = 1$, $\lim\limits_{x \to +\infty} f(x) = 0$. The solution of Problem 1 is contained in the following theorem.

4. THEOREM 2

Denote $c = \sup \left\{ x \in \mathbb{R}: \text{mes}(\{F_0 - F_1 \ge x\}) \ge \dfrac{1}{M} \right\}$.

If $\text{mes}(\{F_0 \ge F_1\}) \ge \dfrac{1}{M}$, then

$$\sup_{f \in L_M} \Phi(f) = M \int_E (F_0 - F_1)(x)dx \quad ,$$

where E is a set defined by the relations

$$\text{mes}E = \frac{1}{M}, \quad \{F_0 - F_1 > c\} \subset E \subset \{F_0 - F_1 \ge c\} \quad . \tag{3}$$

If $\text{mes}(\{F_0 \ge F_1\}) < \dfrac{1}{M}$, then

$$\sup_{f \in L_M} \Phi(f) = M \int_{\{F_0 > F\}} (F_0 - F_1)dx \quad .$$

In (and only in) the case $\text{mes}(\{F_0 \ge F_1\}) \ge \dfrac{1}{M}$, $\sup \Phi(f)$ over $f \in L_M$ is attained, and the optimal function f^* is given by Equation 2 with a set E subject to relations 3.

It follows from the above formulas that the Lipschitz theoretical bound increases with respect to D and decreases with respect to m.

Now consider Problem 2. In Case 1 for "almost all" pairs of measures μ_0, μ_1, the optimal irradiation scheme D^* appears to be uniform, i.e., $D_1^* = D_2^* = \ldots = D_n^*$, and we have the following *a priori* estimate for the number of treatments:

$$n^* \leq \left(\frac{\overline{D}^{m+1} \int_{R_+} t^{m+1} d(\mu_0 - \mu_1)_-}{(m+1)! \overline{\Phi}} \right)^{1/m}, \tag{4}$$

where $\overline{D}$ is the maximal tolerant dose, $\overline{\Phi} = \max\{\Phi(D): 0 < D \leq \overline{D}\}$ is the greatest efficiency in the case of single exposure, and μ_- denotes the negative variation of measure μ. Note also that the value Φ in 4 may be replaced by every greater efficiency of a multifractional scheme with the total dose not exceeding $\overline{D}$ provided that such efficiency is known. Calculations show that both the optimal (equal) fractions D_i^* and the corresponding optimal efficiency Φ^* increase with respect to m.

In Case 3 solution to Problem 2 depends crucially on m. For m = 0, any fractionation of a total dose D provides the same efficiency as the single exposure to the dose D. Every optimal multifractional scheme is uniform only for m = 1, and the number of fractions can be estimated by means of 4 where m = 1 and $\overline{D}$ is replaced by D. For m ≥ 2, complete solution to the problem of optimal partition of a given total dose D into two parts is obtained. Iterating this process (i.e., applying it repeatedly to every fraction obtained in previous steps), we construct the diadic fractionated irradiation schemes with the efficiency approximating the optimal one.

Now turn to the solution of Problem 2 in Case 4. It can be easily seen that, leaving beyond trivial cases, the restriction $\Phi_0(\mathcal{D}) \geq \gamma$ may be substituted by $\Phi_0(\mathcal{D}) = \gamma$. Hence, the optimal total dose and the optimal efficiency decrease with respect to γ. Numerical experiments reveal uniformity of the optimal irradiation scheme, i.e., $D_1^* = D_2^* = \ldots = D_{n^*}^*$.

As our simulation study shows, the optimal regimen retains (at least in some situations) a tendency to be uniform with respect to the value of dose per fraction even in the presence of transient processes of radiation damage formation and postirradiation recovery of normal and neoplastic cells.

Appendix 1: The Relationship Between Sets $\bigcup_{M>0} L_M$ and W

To prove that the set $\bigcup_{M>0} L_M$ is dense in W, denote by A the class of all absolutely continuous functions F on $\mathbb{R}_+$ vanishing at infinity, i.e., of functions of the form

$$F(x) = \int_x^\infty f(t)dt, \quad \text{where } f \in L^1(\mathbb{R}_+) \quad .$$

Clearly, $f = -F'$ a.e. with respect to the Lebesgue measure. We supply the linear space A with the norm

$$\|F\|_A = \int_0^\infty |f(t)|dt \quad .$$

Class W is a subset in A characterized by the properties $f \geq 0$ a.e. and $\int_0^\infty f(t)dt = 1$.

Let $F \in W$. Choose $M_0 > 0$ such that $\text{mes}(\{f \leq M_0\}) > 0$. For $M > M_0$, define

$$\lambda_M = \int_{\{f \leq M\}} f(t)dt, \quad f_M = \lambda_M^{-1} f \chi_{\{f \leq M\}}, \quad \text{and } F_M(x) = \int_x^\infty f_M(t)dt \quad .$$

Then $f_M \geq 0$ a.e., and $\int_0^\infty f_M(t)dt = 1$. Hence, $F_M \in W$. Further,

117

$$\lambda_M \to \int_0^\infty f(t)dt = 1 \quad \text{as} \quad M \to +\infty \quad .$$

Therefore $F_M \in L_{M/\lambda_M} \subset \underset{M>0}{U} L_M$. Finally, we have

$$\|F_M - F\|_A = \int_0^\infty |f_M(t) - f(t)|dt = \int_{\{f>M\}} f(t)dt +$$

$$(\lambda_M^{-1} - 1) \int_{\{f \le M\}} f(t)dt = 1 - \lambda_M + (\lambda_M^{-1} - 1)\lambda_M =$$

$$2(1 - \lambda_M) \to 0 \quad \text{as} \quad M \to +\infty \quad .$$

Appendix 2: Proof of Relation 3.5, Section 2.3

For $m \in \mathbb{Z}_+$, define the functions

$$g_m(\alpha) = f_m(\alpha m) = e^{-\alpha m} \sum_{k=0}^{m} \frac{(\alpha m)^k}{k!}, \quad \alpha \in \mathbb{R}_+ .$$

What we have to show is the following.

Proposition —

$$\lim_{m \to \infty} g_m(\alpha) = \begin{cases} 1, & 0 \le \alpha < 1 \ ; & \text{(A.1)} \\[2mm] \dfrac{1}{2}, & \alpha = 1 \ ; & \text{(A.2)} \\[2mm] 0, & \alpha > 1 \ . & \text{(A.3)} \end{cases}$$

Proof — Obviously, $0 < g_m(\alpha) \le 1$ for all $\alpha \ge 0$. For $\alpha > 0$, we have

$$g_m'(\alpha) = -m^{m+1}(m!)^{-1}e^{-\alpha m}\alpha^m ,$$

$$g_m''(\alpha) = m^{m+2}(m!)^{-1}e^{-\alpha m}\alpha^{m-1}(\alpha - 1) .$$

So, the function $g_m'(\alpha)$ is nonpositive, decreases for $\alpha \in [0, 1]$, and takes its minimal value $-b_m$ at $\alpha = 1$, where

$$b_m = m^{m+1}(m!)^{-1}e^{-m} .$$

Note also that by the Stirling formula,

$$b_m \leq \sqrt{\frac{m}{2\pi}}, \quad m \in \mathbb{Z}_+ \quad . \qquad (A.4)$$

If $\alpha = 0$, then Relation A.1 is trivial. For $0 < \alpha < 1$, we have

$$1 - g_m(\alpha) = g_m(0) - g_m(\alpha) \leq \alpha \max_{\xi \in [0,\alpha]} |g'_m(\xi)| = \alpha |g'_m(\alpha)| =$$

$$m^{m+1}(m!)^{-1} e^{-\alpha m} \alpha^{m+1} = \alpha b_m e^{m(\alpha - 1)} \alpha^m = \alpha b_m e^{-m\varphi(\alpha)} \quad ,$$

where

$$\varphi(\alpha) = \alpha - \ln\alpha - 1, \quad \alpha > 0 \quad .$$

Thus, by A.4,

$$1 - g_m(\alpha) \leq \sqrt{\frac{m}{2\pi}} \, e^{-m\varphi(\alpha)}, \quad 0 < \alpha < 1 \quad . \qquad (A.5)$$

The function $\varphi(\alpha)$ is nonnegative for all $\alpha > 0$, and its minimal value equal to zero is attained for $\alpha = 1$. Hence, A.1 immediately follows from A.5.

For $\alpha > 1$,

$$g_m(\alpha) = m^m(m!)^{-1} e^{-\alpha m} \alpha^m \left(1 + \sum_{k=1}^{m} m(m - 1) \ldots (m - k + 1)m^{-k}\alpha^{-k} \right) \leq$$

$$m^m(m!)^{-1} e^{-\alpha m} \alpha^m \sum_{k=0}^{m} \alpha^{-k} \leq \frac{b_m}{m} \frac{\alpha}{\alpha - 1} \quad .$$

Thus, via A.4,

$$g_m(\alpha) \leq (2\pi m)^{-1/2} \alpha/(\alpha - 1), \quad \alpha > 1 \quad , \qquad (A.6)$$

and this yields A.3.

To prove A.2, we need a more delicate argument. First, we note that for all $h > 0$,

$$g_m(1 + h) = g_m(1) + \int_0^h g'_m(1 + t)dt = g_m(1) - b_m \int_0^h e^{-m\varphi(1 + t)}dt$$

and similarly

$$g_m(1 - h) = g_m(1) + b_m \int_0^h e^{-m\varphi(1-t)} dt \quad .$$

We have the following inequalities that can be checked in a standard way:

$$\frac{t^2}{2} - \frac{t^3}{3} \leq \varphi(1 + t) \leq \frac{t^2}{2} \leq \varphi(1 - t) \leq \frac{t^2}{2} + \frac{2t^3}{3}, \ 0 \leq t \leq \frac{1}{2} \quad . \quad (A.7)$$

Applying them, we obtain for $0 < h \leq \dfrac{1}{2}$

$$|g_m(1 + h) - 2g_m(1) + g_m(1 - h)| = b_m \int_0^h (e^{-m\varphi(1+t)} - e^{-m\varphi(1-t)}) dt \leq$$

$$\sqrt{\frac{m}{2\pi}} \int_0^h \left(e^{-m\left(\frac{t^2}{2}-\frac{t^3}{3}\right)} - e^{-m\left(\frac{t^2}{2}+\frac{2t^3}{3}\right)} \right) dt \leq$$

$$\sqrt{\frac{m}{2\pi}} \int_0^h e^{-\frac{mt^2}{2}} (e^{mt^3} - 1) dt \leq \sqrt{\frac{m}{2\pi}} (e^{mh^3} - 1) \int_0^h e^{-\frac{mt^2}{2}} dt \leq$$

$$\sqrt{\frac{m}{2\pi}} (e^{mh^3} - 1) \int_0^\infty e^{-\frac{mt^2}{2}} dt = \frac{1}{2} (e^{mh^3} - 1) \quad .$$

Further, by virtue of A.6,

$$g_m(1 + h) \leq (2\pi m)^{-1/2}(1 + h)/h \leq 3/(2\sqrt{2\pi m}\, h), \quad 0 < h \leq \frac{1}{2} \ ,$$

while in view of A.5 and A.7, we have

$$1 - g_m(1 - h) \leq \sqrt{\frac{m}{2\pi}}\, e^{-\frac{mh^2}{2}}, \quad 0 < h \leq \frac{1}{2} \quad .$$

Set $\tau = \sqrt{\dfrac{m}{2}}\, h$. Then for $m \in \mathbb{N}$, the estimates obtained above can be rewritten in the form

$$|g_m(1 + h) - 2g_m(1) + g_m(1 - h)| \leq \frac{1}{2} (e^{2\sqrt{2}\, \tau^3/\sqrt{m}} - 1) \quad , \quad (A.8)$$

$$g_m(1 + h) \leq \frac{3}{4\sqrt{\pi}\, \tau} \quad , \tag{A.9}$$

$$1 - g_m(1 - h) \leq \sqrt{\frac{m}{2\pi}}\, e^{-\tau^2} \quad . \tag{A.10}$$

Suppose $\{\tau_m\}_{m \in \mathbb{N}}$ is a positive sequence with the following properties:

$$\sqrt{m}\, e^{-\tau_m^2} \xrightarrow[m \to \infty]{} 0 \quad , \tag{A.11}$$

$$\frac{\tau_m^3}{\sqrt{m}} \xrightarrow[m \to \infty]{} 0 \quad . \tag{A.12}$$

It follows from A.11 that $\tau_m \to +\infty$ as $m \to \infty$. Together with A.12, this implies

$$h_m : = \tau_m \sqrt{\frac{2}{m}} \xrightarrow[m \to \infty]{} 0 \quad .$$

Inserting in A.8 − A.10 $\tau = \tau_m$, $h = h_m$, we see that

$$g_m(1 + h_m) - 2g_m(1) + g_m(1 - h_m) \to 0 \quad ,$$

$$g_m(1 + h_m) \to 0, \text{ and } g_m(1 - h_m) \to 1$$

as $m \to \infty$. Consequently, $g_m(1) \xrightarrow[m \to \infty]{} \frac{1}{2}$.

To complete the proof, we have to note only that

$$\tau_m = \sqrt{\ln(m + 1)}, \quad m \in \mathbb{N} \quad ,$$

is an example of sequence satisfying A.11 and A.12.

Remark — In another way, the Proposition may be proven by using the central limit theorem.

References

1. **Albright, N. W.** (1989) A Markov formulation of the repair-misrepair model of cell survival, *Radiat. Res.*, 118, 1–20.
2. **Albright, N. W. and Tobias, C. A.** (1985) Extension of the time-dependent repair-misrepair model of cell survival to high LET and multicomponent radiation, in *Neyman-Kiefer Conference, University of California, Berkeley, July 1983*, Wadsworth Press, Belmont, CA, 397–424.
3. **Alper, T.** (1977) Elkind recovery and "sublethal damage": a misleading association? *Br. J. Radiol.*, 50, 459–467.
4. **Alper, T.** (1980) Keynote address: survival curve models, in *Radiation Biology in Cancer Research*, Raven Press, New York, 3–18.
5. **Barendsen, G. W.** (1980) Variation in radiation responses among experimental tumors, in *Radiation Biology in Cancer Research*, Raven Press, New York, 333–343.
6. **Barsukov, V. S. and Malinovsky, O. V.** (1973) Quantitative description of the process of radiation cell inactivation. I. Basic assumptions. Lethal damage formation, *Cytology*, 15, 1152–1159 (in Russian).
7. **Bharucha-Reid, A. T. and Landau, H. G.** (1951) A suggested chain process for radiation damage, *Bull. Math. Biophys.*, 13, 153–163.
8. **Brenner, D. J.** (1990) Track structure, lesion development and cell survival, *Radiat. Res.*, 124 (Suppl.), 29–37.
9. **Clifford, P.** (1972) Nonthreshold models of the survival of bacteria after irradiation, in *Proc. 6th Berkeley Symp. Mathematical Statistics and Probability*, Vol. 4: *Biology and Health*, University of California Press, Berkeley and Los Angeles, 265–286.
10. **Cox, D. R. and Oakes, D.** (1983) *Analysis of Survival Data*, Chapman and Hall, London.
11. **Curtis, S. S.** (1983) Ideas on the unification of radiobiological theories, in *Proc. 8th Symp. Microdosimetry, Julich, FRG, 1982*, Commission of the European Communities, Luxembourg, 527–586.
12. **Curtis, S. S.** (1982) Lawrence Berkeley Laboratory Rep. LBL-15913.
13. **Danzer, H.** (1934) Über einige Wirkungen von Strahlen VII, *Z. Phys.*, 89, 421–434.
14. **Eisen, M.** (1979) Mathematical models in cell biology and cancer chemotherapy, *Lecture Notes in Biomathematics 30*, Springer-Verlag, Berlin.

15. **Fleming, T. R., O'Fallon, J. R., O'Brien, P. C., and Harrington, D. P.** (1980) Modified Kolmogorov-Smirnov test procedures with application to arbitrarily right censored data, *Biometrics*, 36, 607–626.

16. **Fletcher, R.** (1981) *Practical Methods of Optimization, 2: Constrained Optimization*, John Wiley & Sons, New York.

17. **Fowler, J. F.** (1964) Difference in survival curve shapes for formed multi-target and multi-hit models, *Phys. Med. Biol.*, 9, 177–188.

18. **Garrett, W. R. and Payne, M. G.** (1978) Applications of models for cell survival: the fixation time picture, *Radiat. Res.*, 73, 201–211.

19. **Gehan, E. A.** (1965) A generalized Wilcoxon test for comparing arbitrarily single-censored samples, *Biometrika*, 52, 203–223.

20. **Ginzberg, D. M. and Jagger, J.** (1965) Evidence that initial ultraviolet lethal damage in *Escherichia coli* strain 15 TAU is independent of growth phase, *J. Gen. Microbiol.*, 40, 171–184.

21. **Goodhead, D. T.** (1980) Models of radiation inactivation and mutagenesis, in *Radiation Biology in Cancer Research*, Raven Press, New York, 231–247.

22. **Goodhead, D. T., Thacker, J., and Cox, R.** (1978) The conflict between the biological effect of ultrasoft X-rays and microdosimetric measurements and application, in *Proc. 6th Symp. Microdosimetry, Brussels*, Harwood Publishing, London, 829–843.

23. **Green, A. E. S. and Burki, J.** (1974) A note on survival curves with shoulders, *Radiat. Res.*, 60, 536–540.

24. **Hanin, L. G., Rachev, S. T., and Yakovlev, A. Yu.** (1993) On the optimal control of cancer treatment for nonhomogeneous cell populations, *Adv. Appl. Prob.*, 25, 1–23.

25. **Hanin, L. G., Rachev, S. T., Goot, R. E., and Yakovlev, A. Yu.** (1989) Precise upper bounds for the functionals describing the tumor treatment efficiency, *Lecture Notes in Mathematics 1432*, Springer-Verlag, Berlin, 50–67.

26. **Harris, K. P. and Albert, A.** (1991) *Survivorship Analysis for Clinical Studies*, Marcel Dekker, New York.

27. **Haynes, R. H.** (1966) The interpretation of microbial inactivation and recovery phenomena, *Radiat. Res.*, 6 (Suppl.), 1–29.

28. **Haynes, R. H.** (1975) The effect of reparation process on survival curves, in *Cell Survival after Low Doses of Radiation: Theoretical and Clinical Implications*, Proc. 6th L. H. Gray Conf. Institute of Physics, John Wiley & Sons, New York.

29. **Himmelblau, D. M.** (1972) *Applied Nonlinear Programming*, McGraw-Hill, New York.

30. **Hug, O. and Kellerer, A. M.** (1966) *Stochastik der Strahlenwirkung*, Springer-Verlag, Berlin.

31. **Ivanov, V. K.** (1986) *Mathematical Modeling and Optimization in Cancer Radiation Therapy*, Energoatomizdat, Moscow (in Russian).

32. **Jagers, P.** (1975) *Branching Processes with Biological Applications*, John Wiley & Sons, London.

33. **Kalbfleisch, J. D. and Prentice, R. L.** (1980) *The Statistical Analysis of Failure Time Data*, John Wiley & Sons, New York.

34. **Kantorovich, L. V. and Akilov, G. P.** (1982) *Functional Analysis*, 2nd ed., Pergamon Press, New York.
35. **Kappos, A. and Pohlit, W. A.** (1972) A cybernetic model for radiation reactions in living cells. 1. Sparsely ionizing radiations, stationary cells, *Int. J. Radiat. Biol.*, 22, 51–65.
36. **Kapul'tcevich, Yu. G. and Korogodin, V. I.** (1964) *Quantitative Regularities of Cell Radiation Injury*, Atomizdat, Moscow (in Russian).
37. **Kapul'tcevich, Yu. G. and Korogodin, V. I.** (1964) Statistical models of postradiation recovery of cells, *Radiobiologiya*, 4, 349–356 (in Russian).
38. **Kim, J. H., Kim, S.-H., Perez, A. G., and Fried, J.** (1973) Radiosensitivity of confluent density-inhibited cells, *Radiology*, 106, 447–449.
39. **Knolle, H.** (1988) *Cell Kinetic Modelling and the Chemotherapy of Cancer*, *Lecture Notes in Biomathematics 75*, Springer-Verlag, Berlin.
40. **Kosevich, A. M. and Kruglikov, I. L.** (1986) Mathematical model of interphase cell death, *Radiobiologiya*, 26, 79–83 (in Russian).
41. **Kosevich, A. M. and Kruglikov, I. L.** (1987) Mathematical model of interphase cell death. Biophysical justification and generalization, *Radiobiologiya*, 27, 12–15 (in Russian).
42. **Kruglikov, I. L.** (1987) Generalization of some microdosimetric correlations for a random non-Poisson process, *Radiobiologiya*, 27, 836–838 (in Russian).
43. **Kruglikov, I. L.** (1989) Dose-response relationship for a stochastic function of cell reactivity, *Radiobiologiya*, 29, 415–417 (in Russian).
44. **Kruglikov, I. L.** (1990) Survival of the irradiated cells as a solution of the stochastic differential equation. General description, *Radiobiologiya*, 30, 207–210 (in Russian).
45. **Kruglikov, I. L.** (1990) Dose-effect relations for stochastic rates of injury formation in cells, *Radiat. Res.*, 122, 120–125.
46. **Kruglikov, I. L.** (1991) Description of the radiation cell death by the stochastic differential equation, *Radiat. Res.*, 127, 97–100.
47. **Laurie, J. J., Orr, J. S., and Foster, C. J.** (1972) Repair processes and cell survival, *Br. J. Radiol.*, 45, 362–368.
48. **Mantel, N.** (1966) Evaluation of survival data and two new rank order statistics arising in its consideration, *Cancer Chemother. Rep.*, 50, 163–170.
49. **Munro, T. R. and Gilbert, C. W.** (1961) The relation between tumour lethal doses and the radiosensitivity of tumour cells, *Br. J. Radiol.*, 34, 246–251.
50. **Neyman, J. and Puri, P. S.** (1976) A structural model of radiation effects in living cells, *Proc. Natl. Acad. Sci. U.S.A.*, 73, 3360–3363.
51. **Neyman, J. and Puri, P. S.** (1981) A hypothetical stochastic mechanism of radiation effects in single cells, *Proc. R. Stat. Soc. London, Ser. B*, 213, 134–160.
52. **Obaturov, G. M.** (1987) *Biophysical Models of the Radiobiological Effects*, Energoatomizdat, Moscow (in Russian).
53. **Obaturov, G. M., Matveeva, L. A., Tyatte, E. G., and Yas'kova, E. K.** (1982) The stochastic model of formation of chromosome aberrations of cells, *Radiobiologiya*, 20, 803–809 (in Russian).

54. **Oliver, R.** (1964) A comparison of the effects of acute and protracted gamma-irradiation on the growth of seedlings of Vicia Faba. II. Theoretical calculations, *Int. J. Radiat. Biol.*, 8 (5), 475–488.

55. **Opatowski, I.** (1945) Chain processes and their biophysical applications. Part I. General theory, *Bull. Math. Biophys.*, 7, 161–180.

56. **Opatowski, I.** (1946) Chain processes and their biophysical applications. Part II. The effect of recovery, *Bull. Math. Biophys.*, 8, 7–13.

57. **Opatowski, I.** (1946) The probabilistic approach to the effect of radiations and variability of sensitivity, *Bull. Math. Biophys.*, 8, 101–109.

58. **Orr, J. S., Wakerley, S. E., and Stark, J. M.** (1966) A metabolic theory of cell survival curves, *Phys. Med. Biol.*, 1, 103–108.

59. **Ostashevsky, J. Y.** (1989) A model relating cell survival to DNA fragment loss and unrepaired double-strand breaks, *Radiat. Res.*, 118, 437–466.

60. **Payne, M. G. and Garrett, W. R.** (1975) Models for cell survival with low LET radiation, *Radiat. Res.*, 62, 169–179.

61. **Payne, M. G. and Garrett, W. R.** (1975) Some relations between cell survival models having different inactivation mechanisms, *Radiat. Res.*, 62, 388–394.

62. **Pepe, M. S. and Fleming, T. R.** (1989) Weighted Kaplan-Meier statistics: a class of distance tests for censored survival data, *Biometrics*, 45, 497–507.

63. **Phelps, R. R.** (1966) *Lectures on Shoquet Theorems*, Van Nostrand, New York.

64. **Pohlit, W. and Heyder, I. R.** (1981) The shape of dose-survival curves for mammalian cells and repair of potentially lethal damage analyzed by hypertonic treatment, *Radiat. Res.*, 87, 613–634.

65. **Porter, E. H.** (1980) The statistics of dose/curve relationships for irradiated tumours. Parts I and II, *Br. J. Radiol.*, 57(7), 210–227 and 336–345.

66. **Powers, E. L.** (1962) Considerations of survival curves and target theory, *Phys. Med. Biol.*, 7, 3–28.

67. **Pury, P. S.** (1982) A hypothetical stochastic mechanism of radiation effects in single cells: some further thoughts and results, in *Probability Models and Cancer*, Le Cam, L. and Neyman, J., Eds., North-Holland, Amsterdam, 171–187.

68. **Rachev, S. T. and Yakovlev, A. Yu.** (1988) Theoretical bounds for the tumor treatment efficacy, *Syst. Anal. Model. Simul.*, 5, 37–42.

69. **Roesch, W. C.** (1978) Models of the radiation sensitivity of mammalian cells, in *Proc. 3rd Symp. Neutron Dosimetry in Biology and Medicine*, Burger, G. and Ebert, H. G., Eds., Commission of the European Communities, Luxembourg.

70. **Rubanovich, A. V.** (1978) A general model of radiation damage and the shape of the dose-effect curve. I. The number of steps in the damaging effect, *Radiobiologiya*, 18, 246–252 (in Russian).

71. **Rubanovich, A. V.** (1978) A general model of radiation damage and the shape of the dose-effect curve. II. Repair in the course of exposure, *Radiobiologiya*, 18, 366–370 (in Russian).

72. **Sachs, R. K. and Hlatky, L. R.** (1990) Dose-rate dependent stochastic effects in radiation cell-survival models, *Radiat. Environ. Biophys.*, 29, 169–184.
73. **Sachs, R. K., Hlatky, L., Hahnfeldt, P., and Chen, P.-L.,** (1990) Incorporating dose-rate effects in Markov radiation cell-survival models, *Radiat. Res.*, 124, 216–226.
74. **Sinclair, W. K.** (1973) N-ethylmaleimide and the cyclic response to X-rays of synchronous Chinese hamster cells, *Radiat. Res.*, 55, 41–57.
75. **Suit, H. D., Sedlacek, R., Fagundes, L., Goitein, M., and Rothman, K. J.** (1978) Time distributions of recurrences of immunogenic and nonimmunogenic tumors following local irradiation, *Radiat. Res.*, 73, 251–266.
76. **Suit, H. D., Shalek, R. J., and Wette, R.** (1965) Radiation response of C3H mammary carcinoma, in *Cellular Radiation Biology*, Williams and Wilkins, Baltimore, 514–530.
77. **Swan, G. W.** (1981) *Optimization of Human Cancer Radiotherapy, Lecture Notes in Biomathematics 42*, Springer-Verlag, Berlin.
78. **Swan, G. W.** (1984) *Applications of Optimal Control Theory in Biomedicine, Pure and Applied Mathematics: A Series of Monographs and Textbooks 81*, Marcel Dekker, New York.
79. **Swan, G. W.** (1987) Tumor growth models and cancer chemotherapy, in *Cancer Modeling. Statistics: Textbooks and Monographs Series 83*, Marcel Dekker, New York, 91–179.
80. **Swan, G. W.** (1990) Role of optimal control theory in cancer chemotherapy, *Math. Biosci.*, 101, 237–284.
81. **Thames, H. D.** (1985) An "incomplete-repair" model for survival after fractionated and continuous irradiations, *Intern. J. Radiat. Biol.*, 47, 319–339.
82. **Thames, H. D.** (1987) Repair of radiation injury and the time factor in radiotherapy, in *Cancer Modeling. Statistics: Textbooks and Monographs Series 83*, Marcel Dekker, New York, 269–314.
83. **Tobias, C. A.** (1985) The repair-misrepair model in radiobiology: comparison to other models, *Radiat. Res.*, 8 (Suppl.), 77–95.
84. **Tobias, C. A., Blakely, E. A., Ngo, F. Q. H., and Yang, T. C. H.** (1980) The repair-misrepair model of cell survival, in *Radiation Biology in Cancer Research*, Raven Press, New York, 195–230.
85. **Tucker, S. L., Thames, H. D., and Taylor, J. M. G.** (1990) How well is the probability of tumor cure after fractionated irradiation described by Poisson statistics? *Radiat. Res.*, 124, 273–282.
86. **Turner, M. M.** (1975) Some classes of hit-target models, *Math. Biosci.*, 23, 219–235.
87. **Ward, J. F.** (1988) DNA damage produced by ionizing radiation in mammalian cells: indentities, mechanisms of formation, and repairability, *Progress in Nucleic Acid Research Molecular Biology 35*, Cohn, W. and Moldave, K., Eds., Academic Press, New York, 95–125.
88. **Williams, T.** (1969) The distribution of inanimate marks over a non-homogeneous birth-death process, *Biometrika*, 56, 225–227.

89. **Yakovlev, A. Yu. and Yanev, N. M.** (1989) *Transient Processes in Cell Proliferation Kinetics, Lecture Notes in Biomathematics 82,* Springer-Verlag, Berlin.

90. **Yakovlev, A. Yu. and Zorin, A. V.** (1988) *Computer Simulation in Cell Radiobiology, Lecture Notes in Biomathematics 74,* Springer-Verlag, Berlin.

91. **Yanev, N. M. and Yakovlev, A. Yu.** (1985) On the distribution of marks over a proliferating cell population obeying the Bellman-Harris branching process, *Math. Biosci.,* 75, 159–173.

92. **Yang, G. L. and Swenberg, C. E.** (1986) Stochastic models for cells exposed to ionizing radiation, in *Modeling of Biomedical Systems,* Eisenfeld, J. and Witten, M., Eds., Elsevier/North-Holland, Amsterdam, 85–89.

93. **Yang, G. L. and Swenberg, C. E.** (1991) Stochastic modeling of dose-response for single cells in radiation experiments, *Math. Sci.,* 16, 46–65.

94. **Zigliavskij, A.** (1985) *Mathematical Theory of Random Search,* Leningrad University Press, Leningrad (in Russian).

95. **Zimmer, K. G.** (1961) *Studies on Quantitative Radiation Biology,* Hafner, New York.

96. **Zinninger, G. F. and Little, J. B.** (1973) Fractionated radiation response of human cells in stationary and exponential phases of growth, *Radiology,* 108, 423–428.

Subject Index

130   Biomathematical Problems in Optimization of Cancer Radiotherapy

Dose, absorbed, 11, 13, 54
  single optimal, 74, 85, 88, 101
  stochastic inactivating, 11, 14
  tolerant, 70, 116
    maximal, 70, 116
Dose-effect curve, 7, 8, 10, 24, 26, 28
Efficiency,
  of cancer treatment (therapy), 1, 31,
    55, 62
  criterion, 55
  of fractionation procedure, 82
  "multihit" maximal, 75, 95
  optimal, 74, 94, 99, 114, 116
    "real", 74
    stepwise, 101–104
    two-fraction, 94, 95
    unconditional, 104
    "single", 83
  functional, 2, 56, 75, 104, 111
    extremal value, 82, 104, 105
    precise upper bound, 2, 58, 105,
    113
    stability, 104, 105
Estimator, 18
  consistent, 26
Evolution operator, 51
Expectation, 2, 11, 16, 21, 34, 55, 70
Explantation of cells, 7, 30
Exponential phase of cell culture
    growth, 28, 29
Fixation time, 21
Flexible simplex algorithm, 25, 95, 109
Fractionation, 21, 47
  optimal, 81, 82, 86, 94
  problem, 81, 85, 88
  procedure, diadic, 82, 95
  scheme, optimal, 87, 94
  uniform, 82, 99
Function, absolutely continuous, 59,
    60, 62, 114, 117
  analytic, 83
  beta, incomplete, 8
  characteristic, 61, 114
  continuous at a point, 61
  extremal, 71
  gamma, 24
    incomplete, 8, 9
  generating, 17, 35–37
  left continuous, 72, 73
  measurable, 65, 72
  nondecreasing, 11, 115
  nonnegative, 69, 120

optimal, 114, 115
real, 65, 72
Gamma distribution, 12, 24, 108
Heterogeneity of cell population, 1
  of lesions, 7
  spatial, 55
Hit and target hypothesis, 63
  model, 3, 5, 10, 15, 58, 69, 81
  principle, 5
  theory, 5, 6, 9
Identification of model, 28, 30
Integer part of a number, 77, 88, 92
Intensity, 60, 62
Intermediate state, 14, 15
  probability, 15
Kantorovich-Rubinstein distance, 71
Kaplan-Meier estimate, 40, 43, 44
Kinetics of cell population, 20
Kolmogorov equation, forward, 12, 20
Laplace transform, 83
Least squares estimate (LSE), 25
Lesion,
  accumulation, 12
  evolution, 19, 50
  formation, 11, 15, 20, 48
    probability, 13
  interaction, 19, 23, 50
    quadratic, 20
  number, 12, 13, 16, 18, 19, 21–23,
    50, 54, 58, 113
    distribution, 15, 19
    expected, evolution of, 21
  repair rate, 19
  transformation, 21
Lesion, irreparable, 50, 51, 107
  lethal, 19, 22
  misrepaired, 9, 18–20
    lethally, 19
  primary, 7, 19
    radiation-induced, 7, 28
  repairable, 21, 23
  repaired, 20
  secondary, 20
  unrepaired, 9, 50, 57, 75
    critical value distribution, 75
Limit point, 83
Lipschitz bound, 2, 70, 72, 75, 79, 115
  condition, 59, 63, 64, 71, 115
  constant, 62, 63
Log-likelihood, 25
Mark, 15, 16

Milton Keynes UK
Ingram Content Group UK Ltd.
UKHW040051071024
449327UK00019B/473